Acousto-Optical Laser Systems for the Formation of Television Images

Acousto-Optical Laser Systems for the Formation of Television Images

Yu. V. Gulyaev, M. A. Kazaryan
Yu. M. Mokrushin, O. V. Shakin

CISP

CRC Press
Taylor & Francis Group
Boca Raton London New York

CRC Press is an imprint of the
Taylor & Francis Group, an **informa** business

Translated from Russian by V.E. Riecansky

CRC Press
Taylor & Francis Group
6000 Broken Sound Parkway NW, Suite 300
Boca Raton, FL 33487-2742

First issued in paperback 2021

© 2019 by CISP
CRC Press is an imprint of Taylor & Francis Group, an Informa business

No claim to original U.S. Government works

ISBN 13: 978-1-03-223582-0 (pbk)
ISBN 13: 978-1-138-59520-0 (hbk)

DOI: 10.1201/9780429488399

Visit the Taylor & Francis Web site at
http://www.taylorandfrancis.com

and the CRC Press Web site at
http://www.crcpress.com

Contents

Introduction

The creation of devices capable of displaying a large amount of information with high quality of the reproduced image is of practical interest in such areas of science and technology as optical processing of information, recording information on various types of media, the reproduction of television images, communication and others where we are dealing with large flows information in real time.

At present, great efforts of the leading electronic companies of the world are aimed at creating television systems for displaying information using laser light sources. Lasers provide high brightness and color contrast in the image, unattainable for lamps and phosphors. Among the laser sources, pulsed lasers are emitted, which allow the effective nonlinear conversion of radiation to other parts of the visible spectrum and, thus, cover the entire wavelength range available for visual perception. One of the promising methods of real-time imaging for these lasers is the pulse image projection method of an amplitude-modulated ultrasonic line that fills the aperture of the acousto-optical modulator.

In the modulation method under consideration, there is no high-speed scan in a row, and, unlike existing methods that use matrix modulators, there is no discrete structure in the image. It is formed in real time without delay and is better consistent with a consistent way of transmitting information over the communication channel. Image dimensions can easily be transformed without modifying the modulation devices themselves. When recording information on various types of media, it is possible to perform coherent optical processing of this information. The possibility of using fully acousto-optical control devices for the system under consideration makes it possible to reject for a number of problems such mechanical control devices as mirror scanners, polyhedral rotating prisms, matrices, and micromirror lines. In addition, crystal media used for modulators can withstand large average and pulsed laser radiation powers, which makes it possible to use the systems in question for technological purposes.

Despite the fact that the impulse imaging method has been known for a long time, there are many unresolved issues related to the efficiency and quality of image formation with the most effective acousto-optical modulators (AOM) on a paratellurite (TeO_2) crystal with amplitude modulation of ultrasound. The complexity of the problem lies in the fact that acousto-optical diffraction must be considered for an anisotropic gyrotropic medium, which is a TeO_2 crystal, and for an intermediate mode of diffraction of light by sound. The question of the prospects of practical application of this method for displaying full-colour television information on large screens in the high-definition standard is unclear.

Of great importance is the question of optimizing the output parameters of the laser for the purpose of imaging by the pulsed method. A copper vapour laser can be chosen as such a source, which remains one of the most powerful sources of light in the visible region of the spectrum and, in its output characteristics, agrees quite well with the requirements for a pulsed imaging system. In Russia, work is continuing to improve these lasers in the direction of increasing pump efficiency and practical efficiency.

The aim of the authors' research is the development of the theory of acousto-optical interaction for anisotropic crystalline media possessing gyrotropic properties and on its basis the development of a technique for calculating the spatial distribution of the intensity of light radiation on the projection screen during the diffraction of pulsed laser radiation by an amplitude modulated ultrasonic signal in a paratellurite crystal (TeO_2).

The study of the characteristics of an acousto-optical system with a pulsed method of forming a line for displaying and recording information using copper vapor lasers makes it possible to extend the results obtained by the authors to systems using full-colour pulsed solid-state lasers, which are now rapidly developing.

Methods of producing laser projection images

1.1. Development of laser television display devices

At the first stages of its development, the work on the creation of television imaging systems was carried out mainly in two directions. The first direction was based on the idea of controlling an electron beam by means of electromagnetic modulation and deflection devices and further converting the energy of electrons into visible radiation on a luminophore screen. This idea was reflected in the creation of electron-beam tubes, which for a long time were the main elements of television imaging devices [1]. Another direction was the work on the direct use of the energy of light beams to create a television image on the projection screen.

The principle of the development of scanning in electron-beam tubes was directly related to the consistent principle of obtaining and transmitting information over the communication channel. In modern television devices there is an electronic memory per frame, which makes it possible to convert a serial information array into a parallel one and then output the image to a matrix stationary screen. In this case, there is no need for scanning systems in line and frame. Each element of the image on this screen takes its place. Currently, there is a large number of different matrix information display systems built on different physical principles, which differ from each other in the ways of creating light and methods of modulation. Their work is described in sufficient detail in the literature [2]. To create a light image in modern matrix TVs, liquid crystal [3, 4] and plasma panels [5] are used. Displays based on organic light-emitting diodes are developed [6, 7]. The technology of creating a colour image by exposure of luminophores when a modulated electron beam interacts with it is being further developed. As a result of the use of cold cathodes with field emission, it was possible to create a flat matrix display consisting of a triad of RGB-luminophores, which are excited by the field emission current [8]. The control field is only 1.5–2.0 V/μm. The brightness of the indicators reaches several thousand candelas per square meter. Further development of this method is the work on the creation of flat displays with surface emission of electrons [9]. For large collective screens, LED matrixes as well as matrices of cathodoluminescent lamps are widely used. Each of the existing technologies has its advantages and disadvantages. The shortcomings of liquid crystals include insufficiently large viewing angles, the inability to display black and the inertia. The plasma has other problems: burn-out of pixels, smaller brightness compared to LCD, and higher power consumption. A prolonged demonstration of a still image is contraindicated in the plasma panel. Common shortcomings of matrix systems are the discreteness of their structure, the unevenness of the luminescence of individual elements, which is especially pronounced when observing images on large poly-screens.

At the heart of modern devices with projection tubes, for example, Barco Reality 812 (Belgium), lies the principle of optical magnification of the image obtained on the screen of high-brightness phosphor cathode ray tubes with a screen diagonal of 300 mm and with magnetic focusing of the electron beam. The image from the screen of the tube is projected by means of a lens or a mirror-lens lens onto the projection screen. Each of the three RGB-tubes has a high resolution (2000 TV-lines) with a vertical scanning up to 200 Hz. At very high quality of the received image the light flux of such projectors is small (up to 500 lumens), which limits the area of their use in closed halls.

A separate direction, which continues to develop only in Russia, is the work on the creation of television projectors based on semiconductor lasers pumped by an electron beam – quantoscopes [10, 11]. In a quantum tube, the tube screen is made in the form of a plate cut from a semiconductor single crystal with reflective coatings applied to both plane-parallel surfaces. This system plays the role of an optical resonator of a semiconductor laser with electronic excitation. When scanning a single-crystal electron beam, modulated in intensity, it becomes the source of the optical image. Advantages of these devices is a relatively low power consumption (1÷1.5 W/lm)

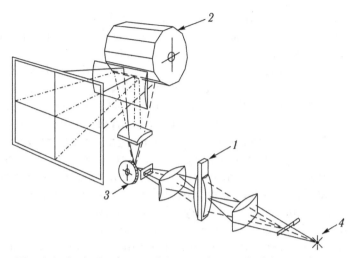

Fig. 1.1. Optical scheme of the Scophony television projector.

with a sufficiently high level of output light flux (up to 3000 lm). Despite the rather high monochromatic radiation, which allows obtaining a high colour contrast, there is no 'specular' structure in the image. Other advantages of modern quantoscopes include a wide range of wavelengths of radiation (460–700 nm), relatively low requirements for projection optics due to the small size of active elements and a small divergence of laser radiation, as well as high resolution (up to 1600 TV lines) and speed, inherent in electron-beam devices. The main disadvantages of the quantoscopes are the presence of high voltage, the possibility of burning out individual active zones and the need for effective cooling of semiconductor wafers of the active element, which limits the possibility of further increase in the dimensions of the plates and the output light flux.

With the use of incoherent light sources, the second trend in the development of television imaging systems found its most complete expression in the television projection system developed in 1938–1939 by the English firm Scophony [12, 13]. Figure 1.1 shows the optical scheme of this device.

As a spatial modulator of light, a liquid ultrasonic cell (*1*) was used in this system. The cell was excited with a travelling ultrasonic wave, which was modulated in amplitude by a video signal. The length of the sound line was chosen equal to the product of the speed of sound in the liquid by the length of the television line. The cell was uniformly illuminated by a light beam from a powerful mercury lamp or an arc discharge (*4*) directed parallel to the front of the elastic wave. After the passage of light through the cell, a picture of the diffraction of light on a volume phase lattice was observed, which was caused by changes in the refractive index of the cell medium due to the elastic-optic effect in the liquid. Using a projection lens from diffracted light beams, the distribution of the light field was formed on the screen, the intensity of which was varied according to the law of amplitude modulation in one coordinate. This image moved around the screen at a speed proportional to the speed of sound in the cell. To compensate for this movement, the light beams were reflected from an additional rotating polyhedral mirror drum (*3*). As a result, the line image became stationary. The consecutive deviation of the amplitude-modulated light lines from another coordinate was realized as a result of the reflection of light from the second mirror drum (*2*) whose rotation speed was synchronized with the frame rate. With the use of a powerful arc lamp, this system made it possible to obtain a television image measuring 4.5×3.6 m^2. With the frequency band of the video channel 5 MHz, the number of elements allowed in the TV line was 250 at 405 lines in the frame. Despite its shortcomings: low resolution, inefficient use of the energy of light radiation, the presence of mechanically controlled elements of light deflection, the creation of this system was a great technical achievement for its time, which was further developed with the advent of powerful sources of coherent radiation-optical quantum generators.

One of the first works on the implementation of a television system for displaying information using a laser was the work of Korpel et al. [14], in which the principle of operation of the projector by Scophony was practically repeated. Figure 1.2 shows the optical scheme of this device. The continuous beam of the He–Ne laser was used as the light beam. The modulator was an ultrasonic cell on water (AOM), in which an acoustic wave with a frequency $f = 41.5$ MHz was excited. Stopping the line on the screen was carried out using an acousto-optical water deflector (ADP), which deflected the modulated light beam incident on it in the direction opposite to where the picture moved on the screen.

The scan on the frame was carried out using a mirror electromagnetic galvanometer. With the frequency band of the video channel 3.15 MHz, this system allowed to obtain 200 elements of resolution per line. The acousto-optical modulators on water used in the first works, because of the large acoustic attenuation (1.5 dB/cm at $f = 30$ MHz), operated at rather low ultrasound frequencies, which

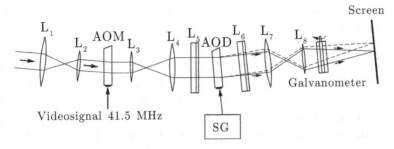

Fig. 1.2. Optical scheme of Korpel's acousto-optical TV-installation. L_1–L_8 – lenses; AOM – acousto-optical modulator on water: AOD – acousto=optical water deflector, SG – sweep generator.

did not allow obtaining a large band of the modulating frequencies of the video channel. In addition, when working with high power ultrasonic signal, the life of such cells was limited.

In the 1970s, in the field of creating laser television display devices, some progress has been made, due to a number of factors. First, continuous ion lasers were created on inert gases: argon (λ_1 = 476.5 nm, λ_2 = 514.5 nm, λ_3 = 488.0 nm) and krypton (λ_4 = 647.1 nm), which could give single-mode output radiation of several watts. Calculations show [15] that when using mixed radiation of these lasers at wavelengths λ_1, λ_2 and λ_4 with output radiation powers, respectively, in the ratio of 1.16; 1; 2.667, one can get a stream of white light. In this case, the radiation at these wavelengths of waves can be used as the primary colours of a full-colour television system. Secondly, new high-efficiency acoustic crystals (TeO_2, NaBi $(MoO4)_2$, α-HJO_3, Hg_2Cl_2, etc.) were synthesized, and broadband piezoelectric transducers were developed to excite elastic waves, which significantly increased the efficiency of laser beam control. Successes were also achieved in the field of creating electro-optical light modulators.

In the first TV-devices with laser beam scanning, continuous inert gas lasers were used: argon, krypton and neon. The peak of development of such systems falls on the 70–80s. At that time, a lot of research work on laser projection devices was published. In such devices, the radiation from an argon or krypton laser is modulated in intensity by acousto-optical [16, 17] or electro-optical [18–21] modulators. It deviates sequentially along the line and frame with the help of acousto-optical deflectors (AOD) [17, 20–23] or optical-mechanical scanning systems, which are rotating mirror drums [18,

24, 25], polyhedral prisms [20], as well as galvanometers [20, 25] and bimorph elements [19].

The main parameters characterizing the operation of the information display system with continuous scanning of the light beam are: the number of solvable elements per line, determined by the product of the bandwidth of the modulating frequencies of the video channel by the duration of the line carrying information, as well as the number of soluble lines in the frame. These parameters are determined by the characteristics of the spatial modulators used in the system and the deflectors of the light radiation. The choice of methods for modulating and deflecting the light beam depends on the requirements imposed on a specific display system or information recording. In the region of low modulating frequencies (≤ 10 MHz), the acousto-optical modulators (AOM) outperform in their parameters the electro-optical modulators (EOM) [26]. The electro-optical modulators have advantages at higher modulation frequencies. With the help of AOM, it is possible to obtain a contrast transfer coefficient at low frequencies that is higher than with EOM.

Opto-mechanical deflectors were used in those cases when it is required to obtain a colour image with low light losses and high resolution in a row and a frame. Since there is no dispersion in such systems, there is no problem with colouring them on the screen. Their shortcomings are inertia, the complexity of providing the necessary accuracy and stability of deflection of light beams, sensitivity to mechanical vibrations. Eliminating these shortcomings is a difficult technical task. In [25], complex automatic systems were used for this, as well as high-speed synchronous motors with magnetic and gas-dynamic suspensions. The acousto-optical deflectors were successfully used in less expensive systems for obtaining a single-colour image [20]. They attract attention by the simple control of the position of the light beam, small dimensions and the absence of mechanically moving parts.

The highest achievement in the field of creating laser television systems for displaying information with continuous lasers was a colour reproducing device developed by Japanese firms NHK and Hitachi for high-definition television [15, 25]. This device allowed to display 1125 lines along the height of the image and had a bandwidth of the video signal path to 30 MHz. With the total power of laser radiation sources of 6 W (4 W – argon laser, 2 W – krypton), the system made it possible to obtain a high-quality image on a 3 m^2 screen. With a screen gain of 4, its brightness was 37.6 cd/m^2. In this system, the power of lasers alone, consumed from the network, was 12 kW, which indicates a low coefficient of efficiency (0.05%) of the laser light sources used. For this reason, these devices have not been widely used.

One of the weakest links in the mechanical system of scans of a television image with continuous light sources is a high-speed rotating multi-faceted mirror drum that forms a string. The speed of rotation of these drums reaches 30÷60 thousand rpm. The quality of manufacturing of these devices is very high. So, for example, for a high-definition television system with a number of solvable elements in a line equal to 1500 [18], the requirements for the accuracy of manufacturing the mirror faces of the drum were as follows:

a) the maximum error in the angular position of the two faces

$$\leq \pm 16 \text{ arc minutes,}$$

b) slope of the surface of the face with respect to the axis of rotation

$$\leq \pm 3.5 \text{ arc minutes,}$$

c) flatness of the surface ≤ 0.05 µm.

Large difficulties in the formation of a uniform linear scan of the image along the line also arise when acousto-optical deflectors of continuous laser radiation are used [15].

The task of eliminating high-speed scanning has always been one of the important tasks facing the creators of light display devices. In the late 30s of the last century, Scophony employees proposed a method of forming a television line on the screen, called the 'wave gap' method, considered as one of the possible options for building a projection system without a high-speed mirror drum. In [27], to stop the motion of the image on the screen, it was suggested to apply a second acoustic cell on the water, in which short ultrasonic pulses, which follow with a repetition rate of the lines, are excited. Light from a continuous source, diffracted by an ultrasonic pulse in the first cell, was projected by the lens onto the aperture of the second modulator in the form of a light spot moving with the speed of sound propagation in the material of the cell. As a result of repeated diffraction of light on the amplitude-modulated video signal by an ultrasonic wave traveling toward the light spot moving across the cell, a picture of the television line was sequentially recorded on the screen. The time of forming the line was equal to half the repetition period of the line pulses, and the scan itself was linear with a high degree of accuracy due to the constant speed of sound propagation in the cell material.

In the installation, it was also supposed to perform pulsed modulation of the light source by means of a Kerr cell [28]. The modulation frequency should be equal to the frequency of the lines, and the cell dimensions should be equal to the product of the speed of sound in the material of the sound line for a time corresponding to the length of the television line. In this case, each light pulse after diffraction on an ultrasonic wave and passing through an optical projection system had to produce a line image on the screen. To obtain a sharp image, the duration of the light pulse must be less than or equal to the travel time of one solvable image element in the modulator. For such a system, the total number of decidable elements in a row can be approximately determined by the expression

$$N_c \simeq \frac{T_c}{\tau + \tau_0} \simeq \frac{T_c \cdot \Delta f_0}{1 + \tau_0 \cdot \Delta f_0}, \qquad (1.1)$$

where T_c is the duration of the line carrying information; τ is the duration of one solvable element, τ_0 is the duration of the light pulse, and Δf_0 is the frequency band of the modulating signal.

It is clear from (1.1) that to obtain 300 solvable elements in the line at $T_c = 60$ μs, $\Delta f_0 = 6$ MHz, it is necessary that $\tau_0 \leq 30$ ns. The frequency of repetition of light pulses should equal the horizontal scanning frequency (15.625 kHz for the SECAM system) at a high average power and radiation direction. Such radiation parameters can not be obtained using incoherent lamp light sources.

In 1974, it was proposed to implement an acousto-optical system with a pulse projection of a television image based on a laser [29].

Figure 1.3 shows the optical scheme of this device.

In the patent it was proposed to use a Nd:YAG laser with frequency conversion to the second harmonic in the Q-switching mode, operating at a repetition frequency equal to the frequency of the television lines. The transition to the visible wavelength range was supposed to be carried out by converting the frequency of the radiation into a second harmonic in a non-linear crystal of lithium iodide. The scanning of the frame in such a device was to be carried out using a mirror drum. As an acousto-optical interaction medium, it was proposed to use an α-HJO$_3$ crystal with a ultrasound propagation

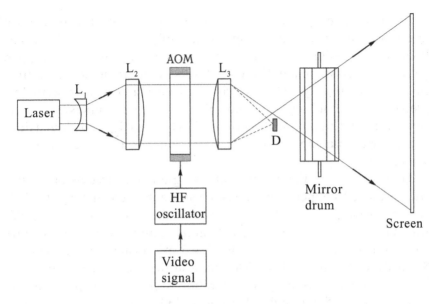

Fig. 1.3. Optical scheme of an acousto-optical television system with a pulsed laser.

velocity of 2.44 · 10^5 cm/s, with a much smaller attenuation coefficient than water. At this speed, the length of the crystal to «write» the entire television line should be 15.5 cm. It was pointed out in [29] that, having radiation at three wavelengths λ_c, λ_s, λ_k in the appropriate power proportions, it is possible to construct a colour reproducing TV image device using one acousto-optical spatial light modulator. For this, it is necessary to excite in the modulator three ultrasonic waves at frequencies f_c, f_3, f_c, satisfying the condition:

$$\lambda_c \cdot f_c = \lambda_3 \cdot f_3 = \lambda_k \cdot f_k = 2 \cdot \sin \theta_B,\qquad (1.2)$$

where θ_B is the angle of incidence of the light beam per modulator equal to the Bragg angle [30].

Under this condition, the images at these wavelengths will be aligned. The author of [29] made an attempt to implement the proposed device in practice, but the technical possibilities of that time: the discrepancy between the parameters of pulsed radiation and the requirements of the impulse projection method, and the absence of suitable acousto-optical crystals, prevented it from being realized. So the Nd:YAG laser operated at 300 Hz and emitted light pulses of 0.2 μs duration, and the length of the acousto-optical cell was only 1/10 of the required one.

Further development of the pulsed method of forming a TV image was achieved with the appearance of powerful pulsed-periodic lasers of the visible wavelength range: metal vapour lasers (copper, gold, etc.), Q-switched Nd:YAG lasers with frequency doubling. In addition, highly effective acousto-optical crystals with low propagation velocities of sound vibrations were synthesized along certain crystallographic directions (TeO_2, Hg_2Cl_2) and dimensions that allowed placing a standard television line (T_c = 52 μs) in the sound line.

Good results on the creation of a single-colour television system with a pulsed copper vapour laser were achieved in England [31] and in Russia [32, 33]. The prototype of the laser installation [195] was successfully tested when demonstrating TV images in city conditions on a screen measuring 4 × 5 m^2, located at a distance of 80 m from the installation. The main drawback of the system was its poor colour, caused by the operation of a copper vapour laser at wavelengths of 510.6 and 578.2 nm. In [34, 35], a system operating on the same principle as the system considered in Refs. [31–33] is described, in which the pulsed Nd:YAG laser with a lamp pumping operating in the Q-switching mode with transformation of radiation into the second harmonic was used. The average output power of this laser at wavelength λ_1 = 532 nm was 16 W. The emission of three such lasers was used to obtain the three main RGB (red, green, blue) colours of a full-colour laser TV image. For a green colour, a Nd:YAG laser with λ_1 = 532 nm and P_1 = 16 W was used. To obtain a red colour, a dye laser was pumped from a second YAG:Nd laser with frequency doubling (λ_2 = 615 nm, P_2 = 15 W). The blue colour was obtained in a laser on Al_2O_3:Ti^{3+} with frequency doubling and pumping from a third Nd:YAG laser with frequency doubling (λ_3 = 450 nm, P_3 = 6 W). This system worked in the standard NTSC and generated a colour TV image on the screen measuring 3.6 × 4.8 m^2. The duration of laser pulses was 100 ns, which made it impossible to obtain a high resolution in the TV projector on a line. [36] reported an increase in the average output power of a blue laser to 7 W (τ_0 = 80 ns, f_{pot} = 17 kHz) at wavelengths of 430 ÷ 460 nm in an Al_2O_3:Ti^{3+} laser with an intracavity frequency doubling on the crystal (VBO), which was pumped by the second harmonic of a Q-switched neodymium laser.

A further breakthrough in the development of laser projection systems occurred in the late 20th and early 21st century. This was due to the development of new fairly powerful laser radiation sources

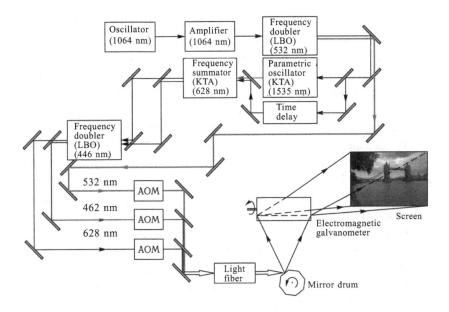

Fig. 1.4. Functional diagram of LDT laser projector.

operating in the green, blue and red spectral regions. All these sources are developed on the basis of non-linear conversion of the infrared wavelength of solid-state or semiconductor lasers into the visible wavelength range. This is most efficient in the pulsed mode of operation of lasers using non-linear optical single crystals or crystals with a regular domain structure (RDS) [37–41].

In 1998, the German company Laser-Display-Technology (LDT) proposed a new technology for creating laser-based projection systems for displaying TV information [42–44], which soon found expression in the mass-produced installations of the German company JENOPTIK.

The laser light source in the installation of this company (Fig. 1.4) is constructed according to the master–amplifier–nonlinear-optical converter scheme. The master oscillator is a $Nd:YVO_4$ crystal laser with LED pumping, operating in the mode locking mode at $\lambda = 1064$ nm and generating 7 ps pulse lengths with a repetition rate of 80 MHz and an average power of 4.5 W. This radiation passes through four cascades of laser amplifiers on active elements of $Nd:YVO_4$ crystals and amplifies to 42 W. Then it enters the non-linear-optical system of radiation transformation, consisting of doublers and frequency combiners on LiB_3O_5 (LBO) and $KTiOAsO_4$ (KTA) crystals, as well as an optical parametric transducer on a KTA crystal.

As a result of this conversion, three laser beams are formed at the wavelengths of 532 nm (P = 6.5 W), 628 nm (P = 7 W), 446 nm (P = 4.8 W). Each of these beams is amplitude-modulated by means of EOM or AOM. All the beams are introduced into one multimode optical fiber, where they are added in power and then fed to a mirror system of deviations in rows and frames, which is a multi-faceted (25 faces) mirror rotating drum and an electromagnetic galvanometer of vertical scanning. Using an optical telescopic lens, a television image is formed on the projection screen. The total output power of the modulated laser radiation in this system is ≈10 W. Because of the very small durations of laser pulses (large values of the intensity of the electric field of light waves), the efficiency of converting the power of infrared radiation into visible light in this system was 40%. In addition, because of the wide frequency spectrum of picosecond light pulses in the image, there is no specular structure, which usually spoils the image in laser projectors. The limiting number of solvable elements in the image for a given system is determined by the ratio of the repetition rate of the light pulses to the frequency of the vertical scan. At a frame rate of 25 Hz, minus the sweepback time, this value is approximately $3 \cdot 10^6$. The radiating fiber end is common for all wavelengths, so the system does not need to combine colours on the screen.

The disadvantages of the system include:

a) drawbacks inherent in the mechanical high-speed mirror system of sweeps used earlier for a system with continuous gas-discharge lasers [15];

b) the complexity of the system design, due to the need to maintain stable temperature conditions for the resonators of the laser system and non-linear optical converters;

c) the high price of a laser projector ($300 000).

For the above-mentioned projection system, work is underway to increase the output power of RGB lasers and to simplify their design. The results of the development of a powerful RGB laser source based on a Yb:YAG thin disk operating in the mode of passive mode locking at a wavelength of 1030 nm and generating 705 fs radiation pulses with a repetition frequency of 57 MHz with the average power of 80 W are presented in Refs [45, 46]. In this setup, it was possible to substantially increase the output radiation power and simplify the design by completely abandoning the laser amplifiers and simplifying the non-linear conversion scheme. For example, crystals with a regular domain structure ($LiTaO_3$) operating at room temperature are

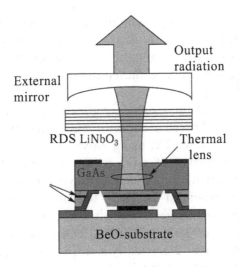

Fig. 1.5. Optical circuit of NECSEL laser.

used in the system. Due to very high values of the intensity of the light field, it was possible to completely abandon resonators with synchronous pumping with parametric transformations. The Output parameters of the developed laser RGB source: average output power at wavelengths $P_G = 23$ W ($\lambda_G = 515$ nm); $P_B = 10.1$ W ($\lambda_B = 450$ nm); $P_R = 8$ W ($\lambda_R = 603$ nm). The coefficient of conversion of infrared radiation into the visible wavelength range in this system is 51%. At present, this system is the most powerful full-colour source of laser radiation.

Another promising technology for the creation of laser television projectors is currently technology based on the use of Novalux RGB lasers [47, 48]. In 2006, at an exhibition in Las Vegas, this company introduced its development of NECSEL semiconductor lasers (Novalux Extended Cavity Surface Emitting Laser). These are semiconductor lasers with a vertical resonator and a radiation output through the side surface (VCSEL) (Fig. 1.5), in the resonator of which a non-linear element with a regular domain structure (RDS) based on MgO:LiNbO$_3$ is built in, which makes it possible to produce an effective intracavity doubling of the fundamental radiation frequency.

The company succeeded in developing RGB lasers emitting in a quasi-continuous mode ($f_{pt} = 500$ kHz, $\tau_0 = 200$ ns) at wavelengths $\lambda_G = 532$ nm, $\lambda_B = 465$ nm, $\lambda_R = 620 \div 635$ nm. The average power radiation of one laser is from 50 to 120 mW with a full efficiency of

$5\div10\%$. Lasers can be easily combined into arrays. The total output power of the lines of 14 individual lasers can reach 1.5 W. Using arrays consisting of a large number of independent laser emitters allows to significantly reduce the contrast of the observed speckle structures in the image.

In January 2008, at the Consumer Electronics Show, Mitsubishi Digital Electronics America officially unveiled the world's first serial rear-projection laser TV, using Novalux lasers. As a modulator of light it used a matrix of silicon micromirrors DMD (Digital Micromirror Device) of the Arasor Company (Australia) which is based on the development of the Texas Instrument Corporation, which created a new type of imager – a digital micromirror device DMD. The matrix consists of 1920×1080 silicon mirrors measuring 4×4 $(\mu m)^2$. Each mirror is electrically controlled and has two stable positions, so the power of the light reflected from the mirrors is controlled by the time duration of the applied voltage. In contrast to modulators on liquid crystals, which can also be used in this projection television, the matrix of micromirrors, according to the developers, gives a significantly higher contrast in the image.

In our opinion, the considered system of forming a TV-image is not devoid of shortcomings. The switching time of micromirrors from one position to another is tens of microseconds, so the dynamic contrast of the image should not be very large. The effect of sticking mirrors is possible. At high light emission powers, there will be problems with the removal of heat from the crystal matrix. The observed effect of changing the light intensity in the image is based on the physiology of the human eye and is of a cumulative nature, which limits the possibility of using these projectors in other tasks, for example, in high-speed display and recording devices operating in real time.

The most promising modulator for the projection systems for displaying TV information with medium-power lasers, both pulsed and continuous, is currently a linear micromechanical modulator consisting of an electrically controlled grating of micromirrors sputtered onto mobile micro substrates of silicon nitride (SiN). This technology modulators is called GLV (Grating Light Valve), it was first proposed in 1992 [49] and was further developed in [50, 51]. The modulator can contain 4096 independently controllable pixels. Each pixel consists of two elements of size (3.7×200 μm^2), one of which is controllable and can be moved by an electric potential within a quarter of the wavelength of the incident light, and the second is immobile. When the voltage is applied to the control element, it is displaced towards the substrate and the light is deflected in the direction of the projection optics. Control of the intensity of reflected light is carried out by the time of finding the light rays in the deflected state. Unlike the matrix, the switching time of one diffractive element in the GLV modulator is only 20 ns. The modulator makes it possible to obtain a one-dimensional amplitude-modulated string consisting of 4096 elements. The scanning of the lines along the frame is supposed to be performed using an electromagnetic galvanometer. According to the developers of the GLV modulator, at a horizontal frequency of 60 Hz one can get a television raster, the number of lines in which 8192.

At present, one can not say that a universal method has been found and a technology has been developed for creating devices for displaying and recording information that satisfies the solution of most practical problems. As a result of the development of science and technology, old technical solutions can be claimed at a new level. One such solution, in our opinion, is the impulse method of forming a TV image using an acousto-optical modulator.

1.2. Pulsed method of forming a television image

Figure 1.6 is an optical scheme explaining the principle of the formation of the image of a television line with the help of a pulsed laser. The central element of the system is the acousto-optical modulator (AOM). In it, a travelling elastic wave is excited with a frequency f, the amplitude of which is modulated by a television

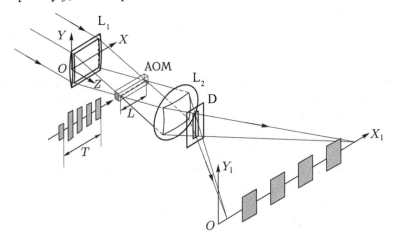

Fig. 1.6. Optical scheme for image formation of a television line with a pulsed laser.

signal in the frequency band Δf_0.

The radiation from the laser with the help of a cylindrical lens L_1 falls in the form of a convergent one coordinate and parallel to the other light beam on the AOM. The angle of incidence of the light wave in the *XOZ* plane (the scattering plane) on the AOM is θ (the angle between the normal to the surface of the modulator coinciding with the *Z* axis and the projection of the wave vector of light onto the *XOZ* plane).

The length of the acoustic tube *L* of the acousto-optical modulator is chosen such that the entire television line is located in it:

$$L = T_c \cdot v \qquad (1.3)$$

where T_c is the length of the line carrying the information, *v* is the speed of sound in the AOM material.

At the moment of filling the modulator's sound line with a sound wave, the laser generates a light pulse and diffraction of light occurs on the sound. In the first order of diffraction, a spectrum of diffracted light waves is formed, corresponding to the spectrum of modulation of the television signal in the line. The lens L_2 builds an image of the line corresponding to this spectrum on the screen. All the extra diffraction orders are filtered by the diaphragm D, located in the Fourier plane of the lens L_2. A TV frame is formed as a result of a successive deviation of the lines by means of a deflector, which can be located between the lens L_2 and the screen directly behind the diaphragm D in the region of the light beam constriction along the coordinate *x*. Thus, for each row carrying information, one image (*x*) is formed on the screen, and on the other (*y*) – the light distribution corresponding to the image of the beam constriction in the focal plane of the lens L_1.

The advantages of this method of image formation include:

a) absence of the high-speed mechanical scanning system in line;

b) high linearity of the formed image along the length of the line, connected with the constancy of the speed of sound in the AOM crystal, while the possible distortions in the image of the line can be related only to the aberrations of the optical system;

c) the possibility of using only solid-state acousto-optical modulators and deflectors to create images;

d) ease of management and input of information into the system;

e) high speed of the system;

e) the ability to control high-power laser beams;

g) low sensitivity to external vibrations;

h) the simplicity of the optical imaging system;

i) the possibility of rapid reshaping of the size of the formed image and its range before installation without loss of the number of solvable elements;

j) the lack of discreteness in the image due to matrix modulators.

To obtain a sharp image of the television line on the screen, the duration of the light pulse should be much shorter than the time for moving one solvable image element in the modulator. The total number of solvable elements in the row N_c, without taking into account the limitations associated with the diffraction effects in the modulator and the optical projection system, can approximately be determined from expression (1.1).

1.3. Pulsed lasers for the television image forming system

1.3.1. Gas lasers

The main requirements that must be taken into account when choosing lasers for a pulsed television image forming system are:

a) operation of the laser in a pulsed mode with a pulse repetition frequency equal to the frequency of the television lines;

b) the short duration of the pulses of laser radiation in accordance with expression (1.1);

c) a sufficiently high average laser radiation power;

d) the possibility of creating a full-colour laser RGB-light source.

Historically, lasers on self-terminating metal atom transitions [52], which until now remain one of the most powerful sources of coherent radiation in the visible wavelength range, were the first lasers that best suited these conditions [53–55]. Generation in these lasers is of a pulsed nature, due to the fact that the inversion of populations in the active medium is achieved for a short time on the transitions between the first resonant and long-lived metastable levels. Excitation of the resonant level occurs due to electron impact in the gas discharge. The characteristic duration of generation is $5 \div 30$ ns. One of the main features of lasers of this type is the high limiting efficiency of the transition, which reaches tens of percent.

The most widespread construction of lasers on self-terminating transitions of metal atoms is a 'self-heating' type laser in which the heating of the metal to the evaporation temperature and excitation of the atoms of the working substance occurs as a result of a longitudinal gas discharge in a tube filled with a buffer gas (usually Ne) hundreds of millimeters of mercury. To create a discharge to the pipe, powerful electric pump pulses are applied. The average power of the applied pulses ensures that the buffer gas is heated up to the required metal vapour density, and the duration of the power pulse front must be much shorter than the lifetime of the atoms at the resonant levels. The limiting repetition rate of the generation pulses is determined by the relaxation time of the metastable levels. For a particular gas-discharge tube, there is some optimal pulse repetition frequency, at which the generation power is maximal [56, 57]. The optimum frequency depends on a number of parameters: the pressure of the buffer gas, its temperature, the diameter of the gas discharge tube, the shape of the gas-discharge channel, and the presence of various impurities of other gases in the pipe. These frequencies usually amount to $8 \div 30$ kHz, corresponding to the frequencies of repetition of television lines. As pulsed hydrogen thyratrons, the characteristics of these, in turn, depend on a number of factors, for example, on the bias voltage, the voltage and the hydrogen generator, the amplitude and duration of the ignition pulse, the switching power, and the switching frequency. The pump efficiency is significantly affected by the duration of the power pulse front applied to the gas discharge tube at the time of opening the thyratron. In practice, the optimization of the operation of lasers of this type is solved experimentally in relation to a specific problem.

To date, lasing has been generated and lasers have been created on the transitions of atoms of various metals, many of the lines being generated lying in the visible wavelength range. Table 1.1 presents the characteristics of the obtained lasing for some lasers on the vapour of metal atoms, which are of greatest interest because of their large practical efficiency.

The best energy characteristics have copper vapour lasers in which the generation occurs at two wavelengths: $\lambda_1 = 510.6$ nm and $\lambda_2 = 578.2$ nm. The lifetimes of the upper working levels corresponding to these two lines are 770 ns and 370 ns. These values are sufficiently large for the known lasers on self-terminating transitions, as a result of which the requirements imposed on the slope of the pump pulse front for these lasers are not large, which allows generating pulses of current through the active element $(50 \div 100$ ns) at sufficiently long durations. For the copper vapour laser, the greatest practical efficiency was obtained, defined as the ratio of the output radiation power on two lines to the power

Table 1.1

Atom	λ, nm	$P_{g.av}$, W	$P_{g.peak}$, kW	τ_0, ns	f_{rep}, kHz	Efficiency, %	T, °C	Ref.
Cu	510.6 578.2	43.5	200	10	20	1,0	1500	[58]
Au	312.2	1.2	13	20	9.1	–		[59]
	627.8	6	22	40	9.1	0.15	1700	[60]
Pb	722.9	4.4	34	5	40	0.2	1000	[61]
Bi	472.2	0.017	–	8	6.25	–	790	[62]
Fe	452.9	–	1	6	1	–	1680	[63]
Ba	1499.9			40				
		12.5	–	25	13.3	0.5	720	[64]
	1130.0							
Mn	534.1	3.5	–	15	5	0.17	1100	[65]

consumed from the network, amounting to 2.6% [66]. The maximum average generation power is 105 W at a frequency of 6.5 kHz with an efficiency of 1% [67]. The maximum frequency of repetition of pulses of 235 kHz with an average radiation power of 0.02 W was achieved, with a power of 2.5 W at a frequency of 100 kHz [68]. The use of gas-discharge tubes with different temperature zones along the length of the channel made it possible to create a multicolour laser in a mixture of copper and gold vapour [69], in which lasing is simultaneously carried out at three wavelengths lying in the visible spectral range: λ_1 = 510.6 nm, λ_2 = 578.2 nm, λ_3 = 627.8 nm. Although the generation of bismuth vapours (λ = 472.2 nm) [62] and iron (λ = 452.9 nm) [63] was obtained, the problem of creating a sufficiently efficient pulsed laser on self-terminating transitions on a blue line with output characteristics close to the characteristics of copper and gold vapour lasers remain open. One of the limiting factors in the development of these lasers is the absence of pumping sources, which make it possible to generate powerful short excitation pulses, the duration of which is equal to nanoseconds. The available experimental data [70, 71] show that with a decrease in the duration of the excitation pulses in a copper vapour laser, the lasing power increases substantially, and the duration of the generation pulses decreases. In this regard, the requirements for the power supply system of these lasers are very high. It can be said that progress in the development of lasers of this class is determined by the successes in the development of effective pumping sources. At present, in addition to the use of traditional pumping schemes with the use of thyratrons, tacitrons, powerful electron tubes, magneto-solid-state pulse formers have been developed [72, 73], which have greater reliability, greater efficiency, and smaller dimensions.

One of the features of lasers on self-terminating transitions of metal atoms is the high gain of the active medium. This property is now widely used to create powerful directional light beams, as well as to enhance the brightness of images of microobjects [74]. On the other hand, the high gain and the short duration of the inversion in the active medium lead to difficulties in the creation of lasers with high radiation directivity. What is needed, for example, for their use in projection systems for displaying and recording information. A decrease in the divergence of the radiation down to diffraction can be achieved by using an unstable resonator with a large magnification coefficient in a laser [75]. In this case, a decrease in divergence by an order of magnitude compared with the divergence of radiation with a stable resonator leads to a decrease in the output power by about 30%. A large output power with a high directivity of the output radiation can be obtained in the system by a master laser-amplifying generator [76]. For effective operation of the system, it is necessary to ensure a strict synchronization of the pump pulses of the first and second lasers. This is due to the fact that the radiation pulse from the generator must fall into the amplifying medium at the time when the inversion in it is maximum at the working wavelength. The strong dependence of the output power on the delay between the pump and generator pumping impulses places high demands on the stability of the delay (a unit of nanoseconds). On the other hand, this property can be used to control the output power of radiation in such a system.

Unfortunately, the choice of sufficiently powerful lasers on self-terminating transitions of metal atoms for information display systems in the visible wavelength range is currently limited to practically only by copper and gold vapour lasers. The other lasers have either low efficiency or emit in the infrared wavelength range, where they can not compete with solid-state laser systems. The best of this class of lasers is a copper vapour laser with a sufficiently high power and good quality of output radiation, which has a limited service life of the active element (2000 hours), large dimensions and low practical efficiency compared to solid-state lasers.

1.3.2. Solid-state lasers

Recently (late 20th – early 21st century), intense development of solid-state pulsed lasers and laser systems with non-linear conversion of infrared radiation into the visible wavelength range has been taking place. Solid-state lasers with optical pumping by laser diodes attract attention with small dimensions, long service life, and high practical efficiency. The controlled pulsed mode of operation of these lasers is realized by intracavity Q-switching using electro-optical or acousto-optical Q-switches. The simplest way of obtaining pulsed radiation in the visible wavelength range is to use the intracavity frequency conversion to the second or third harmonic. Examples of such a transformation include the following published works.

A series of lasers with acousto-optical Q-switching and an intracavity transformation of radiation into the second harmonic have been developed, in which longitudinal optical pumping is performed by diodes through a fiber. In [77], an intracavity transformation was made to the second harmonic of the radiation of an Nd:YVO$_4$ laser at a wavelength of 1342 nm. The non-linear crystal BiB$_3$O$_6$ (BBO) was used for the transformation. The average output power of the radiation is 4.38 W at a wavelength of 671 nm at f_{rep} = 70 kHz and τ_0 = 290 ns with a conversion efficiency of 9.5%.

An intracavity transformation into the second harmonic of an active-element laser on an Nd: GdVO$_4$ crystal was obtained [78]. A non-linear LBO crystal was used for the transformation. The average output radiation power is 6 W at a wavelength of 671 nm at f_{rep} = 47 kHz and τ_0 = 97 ns with a conversion efficiency of 12.8%. The radiation quality factor M^2 = 2.47. The output power fluctuation was 5.8%.

As a result of the intracavity summation of the main radiation and the second harmonic, the output radiation was obtained at a wavelength of 447 nm in an Nd:YAlO$_3$ (Nd:YAP) laser with acousto-optical Q-switching [79]. The second harmonic transformation was performed in an LBO crystal, and the summation occurred in a KTP crystal (KTiOPO$_4$). The average output power at a wavelength of 447 nm was 4.46 W at f_{rep} = 4.6 kHz and τ_0 = 190 ns. Fluctuation of radiation was 3% for 1 hour.

The highest average power of 138 W of output radiation in the second harmonic (λ = 532 nm) of a diode-pumped Nd:YAG laser operating in a Q-switched mode with a frequency of 10 kHz and a pulse duration of 70 ns was obtained in [80]. In this work, the intracavity transformation to the second harmonic was carried out with the help of a non-linear LBO crystal. The efficiency of conversion of pump radiation of laser diodes (λ = 808 nm) to a wavelength of 532 nm was 17.3% with a total laser efficiency of 7.9%. The power density of the laser radiation in the waist was 434 MW/cm^2 with a beam quality factor M^2 = 11. The radiation power fluctuations amounted to 3% during 200 hours of operation. In [81], the output characteristics of this laser were improved. The output power of radiation in 120 W at a frequency of 10 kHz and a pulse duration of 80 ns with a beam quality factor M^2 = 6.2 was obtained.

The relatively long duration of the radiation pulses of these lasers is determined by the use in them of AOM for Q-switching. A shorter generation time can be obtained with the help of EOM. For example, in Ref. 82, a study was reported of a model of a solid-state Nd:YVO$_4$ diode-pumped solid-state laser operating in a Q-switching mode with a repetition frequency f_{rep} = 20 kHz and with an intracavity frequency doubling to a second harmonic on an LBO crystal. In this work, it was possible to obtain a generation time τ_0 = 10 ns for the output pulses due to the use of the Q-switched modulator on a La$_3$Ga$_5$SiO$_{14}$ (LGS) crystal as a Q-switch. The average output power of laser radiation P_{out} at a wavelength of 532 nm was 2.3 W at a maximum optical conversion efficiency of 9.6%. The best radiation quality (M^2 < 2) was obtained at P_{out} = 1.5 W, while the output power fluctuation did not exceed 1.4%.

More complex systems for converting infrared radiation from solid-state lasers are systems using non-linear parametric light conversion. In [83, 84], the source of the main radiation was a diode-pumped solid-state laser (805 nm) on a lithium–yttrium fluorite crystal doped with neodymium (Nd:YLiF$_4$), which was a system consisting of a master oscillator operating in the Q-switching mode, and two cascades of amplification on the same crystal. AOM was used for Q-switching. The output radiation of the main source doubled in frequency in a non-linear LBO crystal. The radiation parameters obtained in this case are the following: the average output power P_{out} = 30 W at a wavelength of 524 nm; repetition frequency f_{rep} = 22.5 kHz; duration of light pulses τ_0 = 35 ns; light quality factor M^2 < 1.2. This radiation was transformed further in a non-linear optical system consisting of an optical parametric oscillator and two frequency doublers on LBO crystals. As a result, at the output of such a system, laser radiation was generated at three wavelengths of

524 nm ($P = 5.8$ W), 628 nm ($P = 6$ W), 449 nm ($P = 3.5$ W), which could be used for a full-colour TV projector with a light output of 4000 lm. The entire laser system consumed 365 watts of power at a full efficiency of 11%.

Compared to the gas lasers on metal vapours, the solid-state lasers have small dimensions, high practical efficiency and a long service life (up to 10 000 hours). Nevertheless, they are currently inferior to certain characteristics of the output radiation, which are of great importance for projection display systems with pulsed lasers. These drawbacks include: a high level of fluctuations in the output power of laser radiation, as well as an insufficiently short duration of light

2

Acousto-optical devices for modulation and deflection of the laser light beam for information display systems

Acousto-optical devices for the modulation and deflection of light beams have been given quite a lot of space in the foreign [85–90] and domestic [91–94] scientific and technical literature. The operation of these devices is based on the phenomenon of light diffraction on periodic perturbations of the refractive index of the medium caused by an ultrasonic wave, and for the purposes of modulation and deviation, the dependence of the intensity and direction of the diffracted light field on the amplitude and frequency of the ultrasonic disturbance is used. Problems solved with the help of acousto-optical devices are extremely diverse. These include optical processing of information [95–98], tunable optical filters [99–102], acousto-optical deployers [103], deflection and modulation of laser radiation [91, 104, 105], broadband delay lines [106], mode synchronizers [107], etc. In each specific case, the design of the ultrasonic cell should optimally correspond to the problem being solved.

2.1. Comparison of light scattering in optically isotropic and anisotropic media

The scattering of light by elastic waves in optical crystals can occur with rotation and without rotating the plane of polarization of the incident light. The scattering of light without rotation of its polarization plane is usually called isotropic, and scattering with rotation of the plane of polarization is anisotropic. If the scattering of light takes place in an optically isotropic medium, this does not lead to any singularities in comparison with the scattering process without rotation of the polarization plane. In scattering in optically anisotropic crystals, the rotation of the plane of light polarization causes significant changes in the usual scattering geometry. This dependence of the angles of incidence and diffraction on the frequencies of the elastic waves is determined by the geometry of the acousto-optical interaction and differs from the usual dependence characteristic of scattering without rotation of the polarization plane.

Investigations of all possible cases of scattering with rotation of the polarization plane appear to be very significant, since they allow us to determine the propagation characteristics of elastic waves necessary to determine the optimal scattering geometry in acousto-optical devices that perform laser radiation control functions: modulators, deflectors, filters.

In the scattering of light by sound, the laws of conservation of energy and momentum must be satisfied [108]:

$$\omega_d = \omega_i \pm \Omega,$$
$$\mathbf{k}_d = \mathbf{k}_i + \mathbf{q},$$

(2.1)

where $\mathbf{k}_i$, $\mathbf{k}_d$ and $\mathbf{q}$ are the wave vectors of the light and sound waves, ω and Ω are the circular frequencies of the light and sound waves, the indices i and d refer to the incident and scattered light, respectively. The wave vector of light is expressed in terms of the wavelength of light and the refractive index: $k = \dfrac{2\pi \cdot n}{\lambda}$. The scattering of light in a crystal can be represented in the form of a vector triangle (Fig. 2.1), which reflects the conservation laws (2.1) for the scattering of light by sound.

The angles of incidence θ_i and the diffraction θ_d of the light waves inside the crystal are measured from the normal to the vector $\mathbf{q}$. Usually the angles are considered positive if they are counted from this normal in the clockwise direction. Nevertheless, for clarity of graphical constructions, in this chapter we will assume that the angles of incidence θ_i and diffraction θ_d in the case of the geometry shown in Fig. 1.2 a, have positive signs.

If the refractive index for the incident and scattered light is the same, then from the conservation laws it follows that $k_d =$

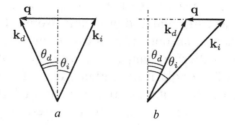

Fig. 2.1. Vector triangles reflecting the law of conservation of momentum for scattering of light by elastic waves in an optically anisotropic crystal without rotating (*a*) and with rotating (*b*) light polarization plane.

$k_i \times$ s $(1 \pm \Omega/\omega_i)$ and, since $\Omega \ll \omega_i$, we get that $k_d \approx k_i = k$. Thus, the vector triangle expressing the law of conservation of momentum is isosceles. From this triangle we get that $\theta_i = \theta_d = \theta_B$ and

$$\sin\theta_B = \frac{q}{2k} = \frac{\lambda_0 \cdot f}{2n \cdot v}, \tag{2.2}$$

where v is the velocity of the elastic waves.

Formula (2.2) corresponds to the usual Bragg condition for the scattering (diffraction) of light in an optically isotropic medium or in an anisotropic medium during the scattering of light without rotation of the plane of polarization. If scattering in an optically anisotropic medium is accompanied by a rotation of the plane of polarization, then the refractive indices for the incident and scattered light turn out to be different, and the vector triangle of the pulses becomes non-isosceles (Fig. 2.1 *b*).

From the triangle shown in Fig. 2.1 *b*, we obtain:

$$\sin\theta_i = \frac{\lambda_0 \cdot f}{2n_i \cdot v}\left[1+\left(\frac{v}{\lambda_0 \cdot f}\right)^2 \cdot (n_i^2 - n_d^2)\right],$$

$$\sin\theta_d = \frac{\lambda_0 \cdot f}{2n_d \cdot v}\left[1-\left(\frac{v}{\lambda_0 \cdot f}\right)^2 \cdot (n_i^2 - n_d^2)\right]. \tag{2.3}$$

It follows from (2.3) that the angles of incidence θ_i and diffraction θ_d differ from each other and depend on the frequency of the elastic waves differently than under the normal conditions [85].

Although the formulas (2.3) are the most general formulas (for $n_i = n_d$ they change to (2.2)). It is convenient to use them only for scattering in the XY plane perpendicular to the optical axis of the crystal. In this case, the refractive indices n_o and n_e do not depend on the direction of the incident and scattered light.

When scattering in an arbitrary plane, the refractive indices n_i and n_d themselves become functions of the angles θ_i and θ_d, and therefore, when considering the features of the scattering geometry, it is more convenient to use the surface of wave vectors whose radius vector determines the magnitude of the wave vector of light propagating in a given direction. In the uniaxial crystals, this surface is a two-sheeted surface consisting of a sphere and an ellipsoid of revolution, which touch each other at two points on the k_z axis [109].

To determine the possible scattering geometry and its dependence on the frequency of the elastic waves, it is necessary to take the cross section of the surface of the wave vectors by the scattering plane and to construct in this section all possible vector triangles expressing the momentum conservation law (2.1) [110].

2.2. Geometry of scattering for various crystallographic planes in uniaxial crystals

Consider the geometry of Bragg diffraction of light on elastic waves for different crystallographic planes, more fully than in Dixon's work [85].

We start with the case when the wave vector of elastic waves is parallel to the Z axis, and the scattering plane is any plane passing through the Z axis. Suppose for definiteness that $n_o > n_e$ (the sphere lies outside the ellipsoid) and that the incident light is ordinary with $k_i > k_d$.

Figure 2.2 schematically shows the dependence of the angles θ_i and θ_d on the frequency of the elastic waves, and also shows the parts of the $k_x k_z$ section of the surface of the wave vectors and three possible vector triangles. The sections are a circle and an ellipse that have a point of tangency on the k_z axis. It follows from Fig. 2.2 that the scattering process can occur from arbitrarily small frequencies of elastic waves, and for $q = 0$, that is, at zero frequency, $\theta_i = -\theta_d = 90°$. As the frequency increases, both angles decrease in absolute value to a certain frequency f_B at which $\theta_i = \theta_{i\ min}$ and $\theta_d = 0$. As the frequency increases further, the angle θ_d becomes positive and both angles increase, reaching, at the frequency f_{max}, the values $\theta_i = \theta_d = 90°$ (backscattering). At a higher frequency, scattering is impossible because of a violation of the momentum conservation law.

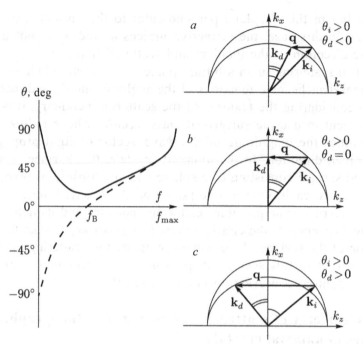

Fig. 2.2. Schematic dependence of the incidence angles θ_i (solid line) and diffraction θ_d (dashed) on the frequency of the elastic waves f, a), b), c) are the cross sections of the wave vector surface by the scattering plane.

Let us consider the case when scattering takes place in the *XY* plane (meaning the crystallographic plane *Y*) and $\mathbf{q}\|Z$. From the vector triangles shown in Fig. 2.1, we see that

$$f_{\hat{a}} = (v / \lambda_0) \cdot (n_0^2 - n_e^2)^{\frac{1}{2}}, \quad f_{max} = 2n_0 v / \lambda_0, \tag{2.4}$$

and the formulas for the angles have the form

$$
\begin{cases}
\sin\theta_i = \dfrac{\lambda_0 \cdot f}{v \cdot n_0} \cdot \dfrac{n_0 \cdot \sqrt{\left(\dfrac{v}{\lambda_0 f}\right)^2 \cdot (n_0^2 - n_e^2)^2 + n_e^2 - n_e^2}}{(n_0^2 - n_e^2)}, \\[4ex]
\sin\theta_d = \pm \left[\dfrac{1 - \left(\dfrac{n_0}{n_e}\right)^2 \cdot \cos^2\theta_i}{1 - \left[\left(\dfrac{n_0}{n_e}\right)^2 - 1\right]\cos^2\theta_i} \right]^{-\frac{1}{2}}.
\end{cases}
\tag{2.5}
$$

The second formula takes the minus sign for $f < f_B$ and the plus sign for $f > f_B$.

Let us consider the case when the wave vector of elastic waves **q** is perpendicular to the optical Z axis, and the scattering plane forms an angle α with the Z axis. For definiteness we set $\mathbf{q} \| X$. The corresponding sections of the surface of the wave vectors by the scattering plane are shown in Fig. 2.3.

From Fig. 2.3 a, it follows that the scattering can begin at arbitrarily small frequencies, and for $f = 0$ the angles are equal, $\theta_i = \theta_d = 0$, and for two non-zero frequencies two scattering geometries can be realized. At one of them, with increasing frequency, the angles increase, remaining positive, and at a certain maximum frequency f_{max} reach the values $\theta_i = \theta_d = +90°$ (backscattering). For another scattering geometry, the angles also increase in absolute value, but here $\theta_d < 0$, and at the frequency f_{max} we have $\theta_i = \theta_d = -90°$. This latter case corresponds to the so-called collinear scattering, in which the wave vectors of the incident and scattered light are parallel.

As follows from (2.4), collinear scattering is impossible for optically isotropic crystals, and also for anisotropic crystals, if the scattering occurs without a rotation of the plane of polarization of light.

Thus, in the case corresponding to Fig. 2.3 a, at frequencies of elastic waves from 0 to f_{0m} two incidence angles and two diffraction angles are possible for each frequency, respectively.

The qualitative course of the dependence of the angles θ_i and θ_d on the frequency of elastic waves is shown in Fig. 2.3 a, and the exact formula for angles has the form

$$\begin{cases} \sin\theta_i = \dfrac{\lambda_0 f}{V}(n_o \pm n_e)^{-1}, \\[4mm] \sin\theta_d = \pm\sin\theta_i \left[1 + \left[\left(\dfrac{n_o}{n_e}\right)^2 - 1\right]\cos^2\theta_i\right]^{-\frac{1}{2}}. \end{cases} \qquad (2.6)$$

Hence we obtain:

$$f_{0m} = \frac{V}{\lambda_0}(n_o - n_e), \quad f_{max} = \frac{V}{\lambda_0}(n_o + n_e). \qquad (2.7)$$

For $\alpha \neq 0$, the circle and ellipse cease to touch each other (Fig. 2.3 b) and their splitting along the k_z axis increases with increasing

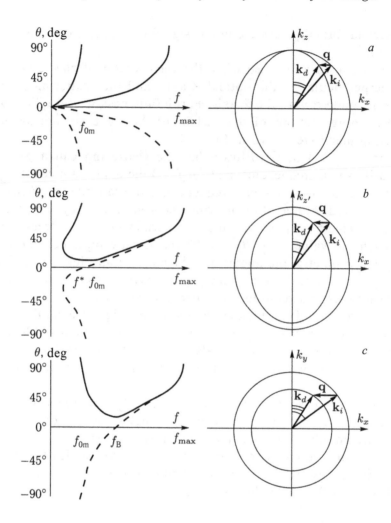

Fig. 2.3. Schematic dependence of the angles of incidence θ_i (solid line) and diffraction θ_d (dashed) on the frequency of elastic waves f, a), b), c) are the cross-section of the surface of wave vectors by the scattering plane. The crystallographic scattering plane is $Z'X$, $\mathbf{q}\|X$. The Z' axis forms an angle α with the Z axis: a) $\alpha = 0$ (scattering plane ZX), b) $0° < \alpha < 45°$, c) $\alpha = 90°$ (scattering plane YX).

α. In this case, as can be seen from the figure, the scattering can begin only with a certain nonzero value f_{min}. In this case, for two frequencies from f_{min} to f_{0m}, as in the case $\alpha = 0$, two scattering geometries are possible, and, at a certain frequency f^*, the angle θ_d is equal to zero. The expressions for the angles in this case have the form

$$\sin\theta_i = \frac{\lambda_0 \cdot f}{v \cdot n_0}\left[\frac{n^2 \pm n_e \cdot \sqrt{\left(\frac{v}{\lambda_0 \cdot f}\right)^2 (n^2 - n_e^2)(n^2 - n_o^2) + n^2}}{n^2 - n_e^2}\right], \tag{2.8}$$

$$\sin\theta_d = \pm\left[\frac{1 - \left(\frac{n_o}{n}\right)^2 \cdot \cos^2\theta_i}{1 + \left[\left(\frac{n_o}{n_e}\right)^2 - \left(\frac{n_o}{n}\right)^2\right]\cos^2\theta_i}\right]^{-\frac{1}{2}}, \tag{2.9}$$

$$f_{\min} = \frac{v}{\lambda_0 n}[(n^2 - n_e^2)\cdot(n_o^2 - n^2)]^{1/2}, \tag{2.10}$$

$$f^* = \frac{v}{\lambda_0}(n_o^2 - n^2)^{1/2},$$

where

$$n^2 = \frac{n_o^2 \cdot n_e^2}{n_o^2 \cdot \sin^2\alpha + n_e^2 \cdot \cos^2\alpha}. \tag{2.11}$$

As the angle α is increased, the frequency $f_{\min}$ increases, and the frequency range $f_{\min} - f_{0m}$, in which there are two scattering geometries, is narrowed. From the condition $f_{\min} = f_{0m}$ we find that two geometries are possible for $\alpha < \text{arctg}\,(n_e/n_o)^{1/2}$. Since $n_e \approx n_o$, then $\alpha \simeq 45°$. For large α, the scattering geometry turns out to be similar to the scattering geometry in the *XY* plane (Fig. 2.3 *c*), and the frequency f^* for $\alpha \to 90°$ smoothly passes into the frequency $f_{\hat{a}} = (V/\lambda_0)\cdot(n_o^2 - n_e^2)^{1/2}$.

When scattering in the *XY* plane, the expressions for the angles are given by the formulas (2.8), (2.9), in which we put $n = n_i = n_0$ and $n = n_d = n_e$. Figure 2.3 *c* shows the angles versus frequency for the cases considered.

Thus, as the transition from the scattering plane *ZX* to the *YX* plane (for **q**||*X*), a smooth transition of the dependences of the angles θ on the frequency f describing the scattering in the *ZX* plane (Fig. 2.3 *b*) occurs to the dependences for the *YX* plane (Fig. 2.3 *c*), where the points f_{0m} and $f_{\max}$ are the 'fixation points' (which do not change when the scattering plane changes).

To complete the analysis of all possible cases, it is necessary to consider scattering in the *XZ'* plane at **q**||*Z'*. In this case, the dependence of θ on f will have some 'intermediate' form if we compare the dependences of Figs. 2.2 and 2.3 *b*, with the 'fixing point' being f_B. The maximum frequency f_{max} will vary from $2n_o V/\lambda_0$ at $\alpha = 0°$ to $(V/\lambda_0) \cdot (n_0 + n_e)$ at $\alpha = 90°$, and the collinear scattering frequency f_0 will vary from 0 at $\alpha = 0°$ to f_{0m} at $\alpha = 90°$ according to the formula

$$f_0 = \frac{v}{\lambda_0} \cdot (n_o - n),$$ (2.12)

where, as before,

$$n = \frac{n_o \cdot n_e}{(n_o^2 \sin^2 \alpha + n_e^2 \cos^2 \alpha)^{1/2}}.$$ (2.13)

It follows from the analysis that the characteristic frequencies of elastic waves are: 1) the backscattering frequencies equal to $f_{max} = 2n_o \cdot v/\lambda_0$ for $\mathbf{q}\|Z$ and $f_{max} = (v/\lambda_0) \cdot (n_o + n_e)$ for $\mathbf{q}\|Z$; 2) the collinear scattering frequency f_0; and 3) the frequency f_B for which $\theta_d = 0$, that is, when the angle of incidence or diffraction is direct. The maximum values of the last two frequencies are, respectively, equal

$$f_{0m} = \frac{v}{\lambda_0} \cdot (n_o - n_e), \quad f_B = \frac{v}{\lambda_0} \cdot (n_o^2 - n_e^2)^{1/2}.$$ (2.14)

The values of these characteristic frequencies for certain uniaxial crystals, which are calculated from formulas (2.7)–(2.14), are presented in Table 2.1. We note that in order for the scattering processes under consideration to take place, the effective photoelastic constant describing the given scattering process [108, 109] must differ from zero. This leads to the fact that some scattering processes turn out to be forbidden. For example, trigonal crystals of point groups D_{3d}, C_{3v}, and D_3 lack collinear scattering for transverse waves propagating along the X axis.

Scattering will, however, occur even with the smallest deviations from the collinear geometry for $\mathbf{q}\|X$ or in the collinear geometry for deviations of $\mathbf{q}$ from X. In this connection, all characteristic frequencies are given for all crystals in the table, even if the given scattering process is impossible, since in the latter case the corresponding characteristic frequency can be regarded as the limiting frequency to which one can experimentally approach arbitrarily closely.

Table 2.1. Characteristic frequencies, scattering (in MHz) for some crystals at $\lambda_0 = 6328$ Å (L, S_f and S_s are longitudinal, fast and slow transverse waves, respectively)

Crystal	Point group	n_o	n_e	Q	Type of waves	f_{0m}	f_B	f_{max}
LiNbO$_3$	C$_{3v}$	2.286	2.200	X	L	890	6428	46 430
					S_f	647	4670	33 744
					S_s	548	3955	28 567
				Z	L	0	7192	52 944
					S	0	3520	25 937
KDP (KH$_2$PO$_4$)	D$_{2d}$	1.5074	1.4668	X	L	384	3290	28 195
					S_f	149	1278	10 945
					S_s	105	900	7703
				Z	L	0	2716	23 575
					S	0	1278	11 094
RDP (RbH$_2$PO$_4$)	D$_{2d}$	1.504	1.475	X	L	223	2263	22 926
					S_f	87	887	8990
					S_s	52	525	5320
				Z	L	0	2044	20 915
					S	0	887	9078
Al$_2$O$_3$	D$_{3d}$	1.765	1.757	X	L	139	2916	61 388
					S_f	86	1792	37 720
					S_s	72	1512	31 835
				Z	L	0	3000	63 293
					S	0	1634	34 474

2.2.1. Anisotropic diffraction in the case of an arbitrary scattering geometry

Following [110], we consider the most general case of scattering, when the scattering plane is rotated around the X axis and forms an angle α with the Z axis, and the wave vector of the sound **q** is directed at an angle β to the X axis (Fig. 2.4).

The system of equations in this case will have the form

$$\begin{cases} n_2 \cos\theta_d - n_o \cos\theta_i = \dfrac{f \cdot \lambda}{v}\sin\beta, \\[2mm] n_o \sin\theta_i + n_2 \sin\theta_d = \dfrac{f \cdot \lambda}{v}\sin\beta, \\[2mm] \dfrac{1}{n_2^2} = \dfrac{\cos^2\theta_d}{n_e^2} + \dfrac{\sin^2\theta_d}{n^2}, \\[2mm] \dfrac{1}{n^2} = \dfrac{\cos^2\alpha}{n_o^2} + \dfrac{\sin^2\alpha}{n_e^2}, \end{cases} \qquad (2.15)$$

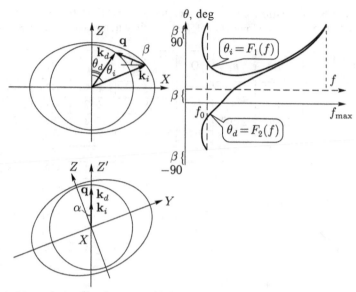

Fig. 2.4. The cross-section of the surface of the wave vectors of an uniaxial crystal by the scattering plane and the dependence of the scattering angles on the frequency of the elastic waves.

from which the following values can be obtained for the diffraction angles:

$$
\begin{cases}
\sin\theta_i = \dfrac{f\cdot\lambda_0}{v}\left(\dfrac{n_e^2\cos\beta\pm n\left[\begin{array}{l}\left(\dfrac{v}{f\cdot\lambda_0}\right)^2(n_o^2-n_e^2)\cdot(n^2-n_e^2)+\\[2mm] +(n^2-n_e^2)\sin^2\beta+2\left(\dfrac{V}{f\cdot\lambda_0}\right)n_o\times\\[2mm] \times(n^2-n_e^2)\sin\beta\cos\theta_i+n_e^2\cdot\cos^2\beta\end{array}\right]}{n_e^2-n^2}\right), & (2.16)\\[12mm]
\sin\theta_d = \pm\left(\dfrac{\dfrac{f\cdot\lambda_0}{v}\cos\beta-n_o\cdot\sin\theta_i}{\sqrt{n_e^2+\left(\dfrac{f\cdot\lambda_0}{v}\cos\beta-n_o\sin\theta_i\right)^2\cdot\left(1-\dfrac{n_e^2}{n^2}\right)}}\right).
\end{cases}
$$

As can be seen from equations (2.15), the characteristic frequencies of the elastic waves are: the backscattering frequency

$$
f_{max} = \frac{V}{\lambda}\cdot(n_o+n') \qquad (2.17)
$$

and the collinear scattering frequency

$$f_0 = \frac{V}{\lambda} \cdot (n_o - n'), \tag{2.18}$$

where

$$n' = \sqrt{\frac{n_o^2 \cdot n^2}{n_o^2 \cdot \cos^2 \beta + n^2 \cdot \sin^2 \beta}}, \quad n = \sqrt{\frac{n_o^2 \cdot n_e^2}{n_e^2 \cdot \cos^2 \alpha + n_o^2 \cdot \sin^2 \alpha}}. \tag{2.19}$$

2.3. Acousto-optical devices and their manufacturing technology

The acousto-optical device consists of an optically transparent medium, a device for exciting elastic waves in it and an electronic control unit. The medium material should be transparent for optical radiation in the operating wavelength range.

To create acoustic waves in an optically transparent medium, a source of elastic oscillations – a piezoelectric plate or a film – is attached to it. The acousto-optical interaction medium may be a liquid or a solid.

Studies were carried out with various substances, but the main attention was paid to solid-state crystalline materials.

The solid-state acousto-optical element consists of a crystalline element with a piezoelectric transducer attached to it. The acousto-optical element is cut out of a certain crystallographic orientation and shape. A piezoelectric transducer attached to it is made of a piezoactive crystal and serves to excite longitudinal or shear vibrations.

The material of the acousto-optical cell can be various crystals, which are chosen on the basis of the requirements imposed on the type of acousto-optical device.

Acousto-optical devices allow controlling the amplitude, phase, frequency and polarization of optical wave radiation. The acousto-optical device for spatial control of the position of the optical beam is called the acousto-optical deflector. The principle of its operation is based on the dependence of the diffraction angle on the frequency of acoustic waves. When the frequency changes, the angle of deflection of the optical radiation changes also.

An acousto-optical device for controlling the amplitude, phase, frequency, and polarization of optical radiation is called an acousto-optical modulator. The principle of its operation is based on the

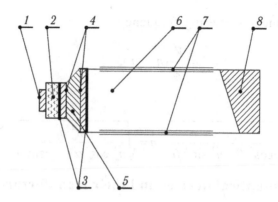

Fig. 2.5. Acousto-optical element with a piezoelectric transducer: *1* – upper electrode; *2* – piezoelectric transducer (lithium niobate crystal $LiNbO_3$); *3* – adhesion layers (chromium Cr, titanium Ti or vanadium V, etc.); *4* – low-resistivity metal layers (copper Cu, gold Au); *5* – connecting layer of a low-melting metal (indium In, tin Sn, etc.); *6* – acousto-optical interaction medium (glasses, crystals); *7* – antireflective coatings of optically polished faces (combination of layers of silicon oxide SiO_2, zirconium oxide ZrO_2, etc.); *8* – absorber of elastic vibrations and heat sink.

dependence of the intensity of diffraction on the amplitude of acoustic oscillations. By controlling the amplitude of the elastic waves, it is possible to change the intensity of the diffracted radiation.

An acousto-optical device for controlling the optical spectrum is called an acousto-optically tunable filter. The principle of its operation is based on the dependence of the diffracted optical wavelength on the frequency of the elastic waves. By controlling the frequency of elastic waves, one can choose the wavelength of light for which the synchronism conditions for acousto-optical interaction are satisfied.

In all types of acousto-optical devices, where traveling acoustic waves are used, the frequency of oscillations of the diffracted radiation differs from the elastic waves incident on the Doppler frequency with which light interacts.

Acousto-optical element. Figure 2.5 schematically shows an acousto-optical element with a piezoelectric transducer attached to it.

The acousto-optical element can be made of various liquid, solid (amorphous and crystalline) materials that are optically transparent in the required operating wavelength range. Solid-state acousto-optical elements can have various shapes to provide acousto-optical interaction conditions.

Piezoelectric transducer. Acousto-optical devices use piezotransducers with acoustic oscillations of the longitudinal or shear directions are used. For the frequency range of 10–1500 MHz, a lithium niobate crystal is usually used as a piezoactive crystal, and ZnO zinc oxide films are used for higher-frequency acousto-optical devices. The limitation of the upper frequency range with a piezoelectric transducer operating on the fundamental harmonic is about 10 GHz. This is due to the very small thickness of the piezotransducer. For example, for a piezoelectric transducer made of a lithium niobate crystal, the thickness for excitation of longitudinal oscillations with a frequency of 1000 MHz should be about 3.5 µm. It is very difficult to grind a plate of lithium niobate to such a thickness, since in grinding the single-domain crystal is destroyed, and its piezoelectric properties are lost. Therefore, for higher frequency acousto-optical devices, zinc oxide films are mainly used.

Control device. The control device consists of a high-frequency generator, an amplifier and a modulator and is connected to the piezoelectric transducer of the acousto-optical cell (AOC) through a device that matches the impedances of the piezoelectric transducer and the output stage of the generator amplifier.

The manufacture of an acousto-optical device consists of a number of operations:

1) the choice of a crystal for an acousto-optical device;
2) calculation of the acousto-optical device depending on its use as a modulator, deflector, filter, etc.;
3) the orientation of the crystals for the production of an acousto-optical cell;
4) manufacturing of crystal elements for an acousto-optical cell and a piezoelectric transducer, attaching a piezoelectric transducer to a crystalline element;
5) grinding the piezoelectric transducer to the required thickness;
6) deposition of the upper electrode;
7) making a bevel on the cell to attach an acoustic absorber;
8) installation of an acousto-optical cell in the housing;
9) matching the wave resistance of the piezoelectric transducer of the acousto-optical cell to the output of the electronic device from which the signal is fed to the piezoelectric transducer (usually at a 50-ohm load).

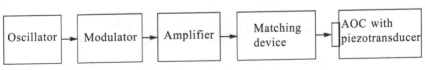

Fig. 2.6. Block diagram of the acousto-optical control device.

2.3.1. Piezotransducer for acousto-optical device

The crystal of the piezoelectric transducer for an acousto-optical device is selected based on the following considerations. For acousto-optical devices, piezo transducers with acoustic oscillations of the longitudinal or shear directions are used. The most widely used piezoelectric converters from the lithium niobate crystal $LiNbO_3$. Bunis of lithium niobate single crystals are usually grown along three directions: along the optical axis, in a direction close to $Y_{+36°}$ and $Y_{+128°}$. $Y_{+36°}$-cut is preferable for excitation of longitudinal oscillations L, since it has a coefficient of electromechanical coupling greater than Z-cut. In Fig. 2.7 presents photographs of the $LiNbO_3$ boules grown along the Z axis by the Czochralski method (provided by VT Gabrielian).

To excite the shear oscillations S, X and $Y_{163°}$ sections are used. X-cut for excitation of shear oscillations of S is preferable, since it has a higher electromechanical coupling coefficient than the $Y_{163°}$-cut.

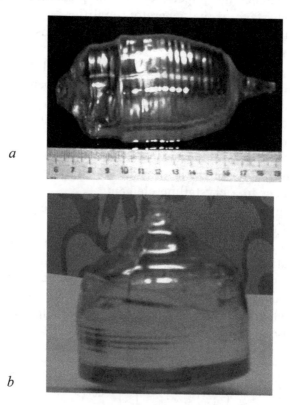

a

b

Fig. 2.7 (*a*, *b*). Photos of $LiNbO_3$ boules grown along the Z axis by the Czochralski method (provided by V.T. Gabrielyan).

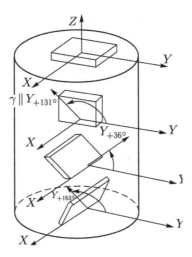

Fig. 2.8. Arrangement of plates of piezoelectric converters in a boule of lithium niobate grown along the Z axis. To excite the longitudinal oscillations of L, $Y_{+36°}$ and Z slices are selected. To excite shear vibrations – X and $Y_{163°}$ sections

In the plate, as shown in Fig. 2.8, at the control of the displacement vector of shear oscillations is located in the direction $Y_{+131,03°}$.

For the manufacture of piezoelectric transducers of the longitudinal type, it is more convenient to use lithium niobate bouilli grown along $Y_{+36°}$, and for shear crystals, crystals grown along the Z axis. Lithium niobate boules grown along the $Y_{+36°}$ direction are oriented by grinding the X plane first, and then the plane $Y_{+36°}$ of the cut.

The base of the crystallographic plane is the cleavage plane $(01\bar{2})$. The cut $Y_{+36°}$ deviates from this plane by an angle equal to $+3°15'$, since the plane $(01\bar{2})$ makes an angle of $32°45'$ with the plane Y or with the plane $(0\bar{1}0)$. The boule with polished mutually perpendicular planes X and $Y_{+36°}$ is sawn on a disk or strip machine to plates about one millimeter in thickness. Then they are polished and cut into the size required for the acousto-optical cell.

To correctly select the slices of lithium niobate crystal plates, it is necessary to accurately determine the positive directions of the Z, Y axes and the X axis in the right coordinate system in accordance with the IRE standard [111]. This can be done with an oscilloscope, observing the electrical pulses when the crystal is touched by the probe connected to the oscilloscope. The faces, from which the positive directions of the Z and Y axes for the lithium niobate crystal come out, are those on which positive electric potentials appear when the crystal is stretched. This is due to the fact that piezomodules for longitudinal deformations d_{22} and d_{33} along the Y and Z axes, respectively, have a positive sign. The orientation of the axes must correspond to the right Cartesian coordinate system.

Shear piezotransducers can be made from a lithium niobate boule grown along the optical axis. To do this, it is necessary to orient and polish the planes Z and X. Plates are cut parallel to the X plane, polished and cut out rectangular plates with a side oriented perpendicular to the direction of shear, that is, in the direction $Y_{+131.03°}$. This direction can be controlled along the crystallographic plane $(10\overline{4})$, which makes an angle of $3.175°$ with the direction of the shear oscillations.

2.3.2. Coefficient of electromechanical coupling

The magnitude of the electromechanical coupling coefficient is the main criterion for choosing an effective piezoelectric transducer for an acousto-optical cell.

The equations of the state of a piezoelectric are obtained from thermodynamic potentials, from which the electromechanical coupling coefficient is determined for a plate or film piezotransducer [112]:

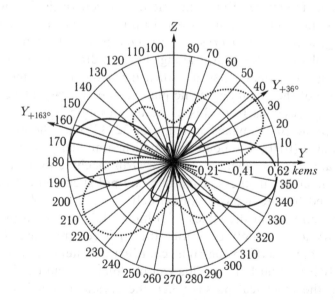

Fig. 2.9. The calculated dependence of the electromechanical coupling coefficient of the lithium niobate crystal for longitudinal L (dashed line) and shear oscillations S (solid line) of Y^*-shaped platelets.

$$K_{ikl} = \frac{d_{ikl}}{\sqrt{s_{klkl}^{E} \cdot \varepsilon_{ii}^{T}}}, \tag{2.20}$$

where s_{klkl}^{E} is the component of the fourth-rank tensor, called the elastic compliance tensor, the superscript E means that this value is obtained at a constant value of the electric field strength E = const; d_{ikl} are components of a third-rank tensor, called the piezomodule tensor; ε_{ii}^{T} are the components of the permittivity tensor at a constant mechanical stress.

It should be taken into account that the coefficient of electromechanical coupling is not a tensor and the transformation formulas for tensors can not be used.

Figure 2.9 shows the dependence of the electromechanical coupling coefficient in a lithium niobate crystal for longitudinal and shear vibrations of Y^*-shape plates.

It can be seen from the figure that with a cutoff angle from the Y axis close to $+36°$, the electromechanical coupling coefficient for longitudinal oscillations reaches a maximum, with no shear oscillations for a homogeneous electric field applied along the cutting direction. For a Z-cut, shear oscillations are also absent. But since the value of the electromechanical coupling coefficient for the $Y_{+36°}$-cut is larger than for the Z-cut, then the $Y_{+36°}$-cut is used to excite the longitudinal oscillations of L. To excite the shear oscillations S, $Y_{+163°}$-cuts are also used because of the absence of longitudinal oscillations in this direction.

The calculation of the piezoelectric transducer made from a lithium niobate crystal is carried out as follows. For longitudinal oscillations

$$K_{22}^{*} = \frac{d_{22}^{*}}{\sqrt{s_{22}^{E*} \cdot \varepsilon_{22}^{S}}}, \tag{2.21}$$

where

$$\begin{aligned} s_{22}^{E*} &= s_{11}^{E} \cdot \cos^4 \alpha + s_{12}^{E} \cdot \sin^4 \alpha - 2s_{14}^{E} \cdot \cos^3 \alpha \cdot \sin \alpha + \\ &\quad + 0.25(s_{44}^{E} + 2s_{13}^{E}) \cdot \sin^2 2\alpha, \\ d_{22}^{*} &= d_{22} \cdot \cos^3 \alpha + (d_{15} + d_{15}) \cdot \cos^2 \alpha \cdot \sin \alpha + d_{33} \cdot \sin^3 \alpha, \\ \varepsilon_{22}^{T*} &= \varepsilon_{11}^{T} \cdot \cos^2 \alpha + \varepsilon_{33}^{T} \cdot \sin^2 \alpha. \end{aligned} \tag{2.22}$$

For shear vibrations

$$K_{22}^* = \frac{d_{24}^*}{\sqrt{s_{44}^{E*} \cdot \varepsilon_{22}^T}}, \qquad (2.23)$$

where

$$
\begin{aligned}
s_{44}^{E*} &= (s_{11}^E + s_{33}^E - 2s_{13}^E - s_{44}^E) \cdot \sin^2 2\alpha + s_{14}^E \cdot \sin 4\alpha + s_{44}^E, \\
\varepsilon_{22}^{T*} &= \varepsilon_{11}^T \cdot \cos^2 \alpha + \varepsilon_{33}^T \cdot \sin^2 \alpha, \qquad (2.24) \\
d_{24}^* &= d_{15} \cdot \cos^3 \alpha + (2d_{33} - 2d_{31} - d_{15}) \cos\alpha \cdot \sin^2 \alpha - 2d_{22} \cos^2 \alpha \cdot \sin \alpha.
\end{aligned}
$$

For the calculation, the following values of piezoelectric modules, dielectric permittivities and elastic compliance coefficients were used:

$$d_{22} = 20.95 \cdot 10^{-12} \, C/N, \quad d_{15} = 65.36 \cdot 10^{-12} \, C/N,$$

$$d_{31} = -1.32 \cdot 10^{-12} \, C/N, \quad d_{33} = 8.27 \cdot 10^{-12} \, C/N,$$

$$\varepsilon_{11}^T = 82.5 \cdot \varepsilon_0, \quad \varepsilon_{33}^T = 28.3 \cdot \varepsilon_0, \quad \varepsilon_0 = 8.8542 \cdot 10^{-12} \, F/m,$$

$$s_{11}^E = 5.86 \cdot 10^{-12} \, m^2/N, \quad s_{33}^E = 5.08 \cdot 10^{-12} \, m^2/N,$$

$$s_{44}^E = 16.88 \cdot 10^{-12} \, m^2/N, \quad s_{11}^E = 13.96 \cdot 10^{-12} \, m^2/N.$$

As can be seen from the calculation and the figure, for some devices it is possible to select the sections of the piezoelectric transducers, which will excite simultaneously both types of oscillations, *L* and *S*. These are $Y_{+12°}$ and $Y_{+137°}$-cuts. The graph also shows that to reduce the level of excitation of the parasitic oscillations we require a high accuracy of the crystallographic orientation of the plates. This accuracy is 5′. If the orientation is incorrect, both types of oscillations will be excited simultaneously. The quality of piezoelectric converters also depends on the degree of single-domain lithium niobate and its homogeneity. This imposes certain requirements on the technology of growing lithium niobate crystals. This refers to a high single-domain, a small spread in the values of piezomodules, the absence of bubbles and inclusions.

The direction of polarization of shear waves for the *X*-cut is calculated on the basis of the following expression:

$$K_{16} = \frac{d_{112}^*}{\sqrt{\varepsilon_{11}^T s_{66}^{E*}}}, \qquad (2.25)$$

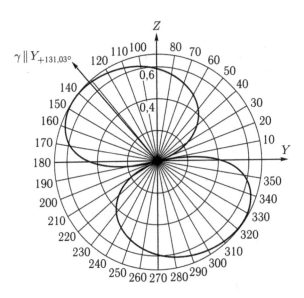

Fig. 2.10. The calculated dependence of the coefficient of electromechanical coupling for a shear piezoelectric transducer made of a lithium niobate crystal X-cut, on the cut-off angle.

where

$$s_{66}^{E*} = 0.5(s_{11}^{E} - s_{12}^{E}) \cdot \cos^2 \alpha + s_{44}^{E} \cdot \sin^2 \alpha + 2s_{14}^{E} \cos \alpha \cdot \sin \alpha,$$

$$d_{112}^{*} = d_{16}^{*} = d_{22} \cdot \cos \alpha - d_{15} \cdot \sin \alpha. \qquad (2.26)$$

Figure 2.10 shows the calculated dependence of the electromechanical coupling coefficient for a shear piezoelectric transducer, made from a lithium niobate crystal of the X cut, on the cut-off angle.

Figure 2.10 shows that the maximum electromechanical coupling coefficient for the shear piezoelectric transducer of the X-cut lithium niobate lies in the direction of $Y_{+131.03°}$, equalling 41.03° to the Z axis.

2.3.3. Attaching a piezoelectric transducer to a crystalline element

Let us consider the peculiarities of the connection of a piezoelectric transducer with a crystalline element in the manufacture of an acousto-optical cell. When welding a plate of a piezoelectric transducer with a crystal element cut out in a certain way from different crystals, there is a risk of cracking either a converter or a crystal. It can be is associated with the difference in the coefficients

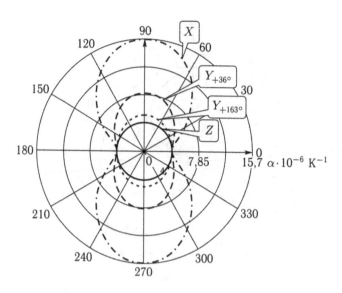

Fig. 2.11. Cross sections of the index surface of the coefficient of thermal expansion of plates of piezoelectric transducers from X -, $Y_{+36°}$-, $Y_{+163°}$ and Z-cuts of lithium niobate.

of thermal expansion of dissimilar welded materials and with the rigidity of welding.

The piezoelectric transducer is attached to the crystalline element by diffusion bonding with a Cr–Cu–In bonding layer. Purified in the vapors of toluene and propanol, the surfaces of the piezoelectric transducer and the sound line are vacuum-deposited with an adhesive layer of chromium (or titanium) with a thickness of 600–800 Å, a copper layer of thickness ~0.2 μm, and an indium layer 1–1.5 μm thick. The thicknesses of the bonding layers are calculated according to a program similar to the program for calculating antireflection coatings.

When attaching the piezoelectric transducer, it is necessary to take into account the thermal expansion coefficients of the piezoelectric transducer plates from lithium niobate and the material of the sound line to which the piezoelectric transducer is attached. Figure 2.11 shows the coefficients of thermal expansion of lithium niobate of various sections used for the manufacture of plates of piezoelectric transducer.

The cross-sections of the index surfaces of the thermal expansion coefficients for piezoactive transducers from lithium niobate (at room temperature) for the Z cut are a circle, and for $Y_{+36°}$, $Y_{+163°}$ and

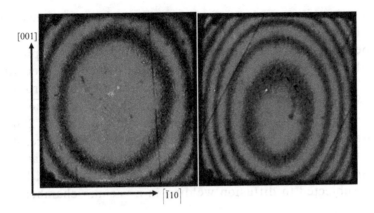

[001]

[Ī10]

Fig. 2.12. Paratellurite crystal in the IT-100 interferometer. On the left, at a temperature of 25°C, on the right it is heated to 45°C. The plane of the figure is parallel to the growth plane of the crystal (110).

X-cuts are eight different sizes. This illustrates the inhomogeneity of the thermal expansion of the plates of piezoelectric transducers and imposes certain limitations when attaching them to the element of the sound line.

If we take, for example, a circular plate of *Z*-cut lithium niobate with a thickness of about 2 mm and a diameter of about 20 mm, polished at a temperature of 25°C, and we place it on the interferometer table, then we see that it is sufficiently flat (has a colour). But if we take in hand and heat the plate by ten to twenty degrees and place it again on the interferometer table, we will see a wavy surface with hexagonal symmetry associated with the crystallographic symmetry of the lithium niobate crystal. A similar phenomenon – thermal distortion of the surface – is also observed for a paratellurite crystal with tetragonal symmetry. Because of the differences in the coefficients of thermal expansion along the optical axis and perpendicular to it [**110**], the interference pattern of the crystal plane polished at room temperature will vary. When heated, the surface becomes convex, and has an elliptical shape elongated along the direction of the optical axis (see Fig. 2.12). Interferograms were obtained on an IT100 interferometer.

It can be seen from the figure that the surface of a crystal treated at room temperature begins to deform upon heating. This complicates the task of thermal-compression welding of piezoelectric transducers. In addition, when the acousto-optical modulator is operating, its operating temperature can be about 60°C and for some devices it should be taken into account, for example, in image processing devices. Correction of the processed plane of the crystal as a function of the operating temperature makes it possible, for example, to decrease the divergence of the acoustic beams, which is especially important for paratellurite crystals with large acoustic anisotropy.

When welding a piezoelectric transducer plate with an acousto-optical element, they are heated to temperatures of the order of 200°C and, naturally, change their dimensions and flatness. Therefore, it is necessary to take into account the thermal properties of the plates of piezoelectric converters and the crystal elements of the sound lines, which are subjected to diffusion welding in a vacuum chamber.

2.3.4. Cleaning of the surfaces of the piezotransducer and sound pipe before the deposition of binding metals

To increase the adhesion of the sputtered metals to attach the piezoelectric transducer in the vacuum chamber, the welded parts are further purified in the discharge of residual gases and argon [113].

It should be taken into account that when bombarded with 500-eV argon ions, the surface of lithium niobate loses oxygen ions and as a result becomes conductive and this layer loses piezoelectric properties. The conductivity effect can be eliminated by prolonged exposure in a humid air atmosphere with preliminary treatment with oxygen plasma. However, the stoichiometry of the crystal is not restored. Therefore, the ion etching of the piezoelectric transducer from lithium niobate must be carried out in an oxygen medium (20% oxygen, and the rest – nitrogen), but the etch rate is almost four times less than in argon etching (4.63 nm/s when etched in argon and 1.35 nm/s for oxygen etching). However, in the case of oxygen etching, the stoichiometry of the congruent lithium niobate and its piezoelectric properties are not affected.

The temperature of the parts to be welded when sputtering chromium and copper is ~200°C, and when indium is sputtered it is ~60°C. After deposition of metals, the piezoelectric transducer is applied to the sound line (without removing them from the vacuum chamber) and pressed down with a force of ~20 kg/cm². In the compressed state, the welded parts are heated to a temperature above the melting point of indium (~157°C) at a rate of 100 deg/hr. The holding time at the maximum temperature is from 7 to 10 min. The cooling rate of the welded parts is 50 deg/h.

a

b

Fig. 2.13. A device for simultaneous welding in vacuum of twenty-eight acousto-optical cells: *a*) in the open and *b*) in the closed state.

All operations are performed in a vacuum without depressurization. Due to this, the welded layer is sufficiently plastic and the difference in the coefficients of thermal expansion of the piezotransducer material and the crystal element of the acousto-optical cell does not lead to cracking of the crystal or converter.

As an example, Fig. 2.13 shows photographs of a device for simultaneous vacuum welding of twenty-eight piezoelectric converters.

2.3.5. Grinding of the piezotransducer after welding to a specified thickness

After diffusion welding, the piezoelectric transducer is ground to a predetermined thickness h, which is determined by formula

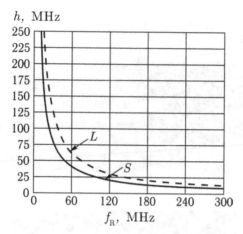

Fig. 2.14. Dependence of the resonance frequency of acoustic oscillations of a piezoelectric transducer made from lithium niobate for excitation of longitudinal *L* (dashed line) and shear oscillations *S* (solid line) in an acousto-optical cell.

$$h = \frac{v}{2 f_R}, \qquad (2.27)$$

where v is the speed of elastic waves in the piezoelectric transducer, and f_R is the resonant frequency. Figure 2.14 shows the dependence of the thickness of the piezoelectric transducer from lithium niobate on the resonant frequency for the *X*-cut (curve *L*) and the $Y_{+36°}$-cut (curve *S*).

This dependence makes it possible to determine which grain size number of the abrasive powder can grind the piezoelectric transducer plate from lithium niobate. The point is that when the surface is grounded, the disturbed layer appears and the plate loses its piezoelectric properties. The thickness of the damaged layer when treated with a free abrasive is approximately two grain diameters of the abrasive.

It is also seen from the figure that the lower the speed of elastic waves, the lower the thickness to which it is necessary to grind the piezoelectric transducer.

For example, at a resonance frequency of 100 MHz, the thickness of the piezoelectric transducer is about 24 μm for the *X*-cut. If we grind to higher frequencies, for example to 200 MHz, then the abrasive should be less than 5–6 μm. With further grinding, it is necessary to choose a finer abrasive with a grain size of 2–3 μm.

In addition, when the surface layer is ground, the surface layer is broken and there is the possibility of scratching, which means that the complete destruction of the piezotransducer depends on the grinding material, its hardness and viscosity. If we grind with a coarser abrasive, then the monodomain is destroyed and the piezotransducer loses its operability. At the same time, under a microscope, it looks like ice in a thaw ('fat').

To excite the longitudinal waves, as can be seen from the graph (Fig. 2.14), the thickness of the piezoelectric transducer from the $LiNbO_3$ $Y_{+36°}$-cut crystal operating at 100 MHz will be approximately 35 μm, which is associated with a higher velocity of longitudinal elastic waves.

The resonant frequency of the piezotransducer is monitored by means of a frequency response meter. The grinding is carried out on powders of various sizes. It is important that the thickness of the transducer plate is at least twice the grain size of the abrasive.

It should be noted that, after grinding, the residual stresses caused by welding are significantly reduced. In addition, after grinding to remove the disturbed layer whose thickness is usually equal to two sizes of abrasive particles, and to bring the piezotransducer to higher resonance frequencies, ion etching in oxygen plasma is used. The etching is carried out through a mask, a part of the transducer plate is etched, onto which the upper electrode is subsequently sprayed. To remove the disturbed layer, it is sufficient to etch from half an hour to an hour. At the same time, a layer of about 5–6 μm is etched away, and this is quite sufficient, since the final grinding is carried out on a diamond abrasive of 2–3 μm.

2.3.6. Orientation and manufacture of acousto-optical cell sound transmission

A wide range of crystals is used as an acousto-optical medium for the interaction of light and elastic waves. At present, the greatest number of acousto-optical devices is made of paratellurite crystals.

The paratellurite TeO_2 crystal exhibits one of the highest values of the acousto-optical quality M_2 when the light travelling around the optical axis interacts with shear acoustic waves propagating and polarized along the [110] direction. It has a large acoustic anisotropy. The velocities of shear elastic waves in different crystallographic directions in paratellurite differ by more than five times (for comparison, in crystalline quartz, only 1.5 times). This feature of the TeO_2 crystal makes it possible to design various acousto-optical devices with an 'unusual' choice of the propagation directions of elastic and optical waves.

To produce high-efficiency acousto-optical devices, work was carried out to investigate the optimum growth conditions for paratellurite single crystals using the Czochralski method. Optimal results were achieved by matching the drawing speed with the growth rate and maintaining the stability of the temperature gradient, as well as by deep purification of the feedstock, improving the design of the growth plant, developing and applying reliable methods for controlling the optical quality of the crystals.

The optimal rate of crystal stretching from the melt in the [**110**] direction is 1.2 mm/h, the seed rotation frequency is 60 rpm, in the absence of thermal and mechanical vibrations. In this case, the tolerance for temperature fluctuation should not exceed ±0.2°C.

Purification of the initial raw material of tellurium dioxide was carried out according to the acid method developed at the Department of Analytical Chemistry of the Dnepropetrovsk State University. The method makes it possible to completely purify TeO_2 from metallic tellurium, copper, iron, selenium, silicic acid and other impurities.

The crystalline perfection of paratellurite is controlled by microscopic examination of etched samples.

As a result of the studies carried out, a technology for growing and controlling samples of high-quality paratellurite single crystals of the required size, more than 40 mm in diameter, has been developed (see Fig. 2.15) [114]. The paratellurite crystal grown by this technology withstands high optical power density and therefore is one of the materials promising to control high-power laser radiation.

Boules of paratellurite crystals are usually grown along the [**110**] direction. Crystals of paratellurite are optically dextrorotatory and levorotatory. There are various methods of crystal orientation [115]. One such technique [116] allows one to achieve the accuracy of orientation of a given crystallographic plane to 10–15″. It is produced using an x-ray goniometer and an optometer. First orient the side and end planes of type (**110**) with a deviation of no more than 2 arc minutes. This accuracy is necessary because of the large elastic anisotropy of the TeO_2 crystal.

To make an acousto-optical cell, the boolean crystal is cut into elements of the required shape. The faces are polished for attaching a piezoelectric transducer and two faces for the passage of optical radiation. Dielectric antireflection films are applied to the optical faces. Layers of silicon oxide SiO_2 and zirconium oxide ZrO_2 are

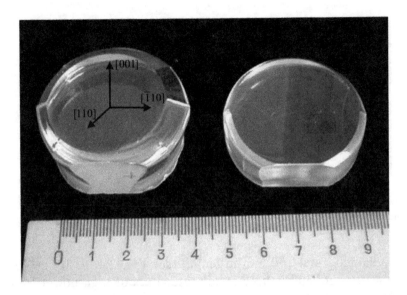

Fig. 2.15. Paratellurite crystals – TeO$_2$, grown by the Czochralski method [113].

deposited on paratellurite. Films of silicon oxide and zirconium have a high hardness and well protect the acousto-optical cell from scratches during surface cleaning.

Molybdate of sodium bismuth NaBi (MoO$_4$)$_2$. In addition to paratellurite, other materials are used for acousto-optical devices. In particular, an effective and technological single crystal of sodium molybdate and bismuth NaBi (MoO$_4$)$_2$ has been developed [117, 118]. It replaces lead molybdate crystals, which, due to internal stresses, are subject to cracking and therefore require additional operations in the manufacturing process of acousto-optical devices. The parameters of NaBi (MoO$_4$)$_2$ are given in Appendix 1.

The sodium molybdate crystal of NaBi (MoO$_4$)$_2$ is almost completely similar to the PbMoO$_4$ lead molybdate crystal, but, in contrast, it withstands significantly higher temperature differences during processing (cutting and grinding). This is due to the lower ionic radii of sodium and bismuth than the ionic radius of lead. In contrast to the bismuth sodium molybdate crystal, some molybdate crystals of lead are destroyed directly in boules. They do not even have time to cut into blanks for sound lines. Elements require additional annealing, which complicates the technology of manufacturing acousto-optical devices.

The bismuth sodium molybdate is well sawn and polished and does not require additional annealing, it grows large enough – 50

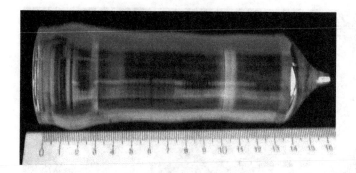

Fig. 2.16. The bismuth sodium molybdate NaBi $(MoO_4)_2$ crystal grown by the Czochralski method [117, 118].

mm in diameter and 150 mm in length (see Fig. 2.16) . It has good radiation strength and is used inside resonators of high-power (more than 100 Watt medium-power) aluminum-garnet lasers operating in the infrared range at a wavelength of about 1 μm. The crystal is grown along the optical axis and the end of the boule and one of the lateral faces of the crystal are oriented.

2.3.7. Attachment of the absorber–heat sink

An absorber of elastic waves, which is also a heat radiator, is attached to the end opposite to the piezoelectric transducer,

It is made of a material having good thermal conductivity and an acoustic impedance close to the acoustic impedance of the material of the sound line. This is due to the fact that the acoustic impedance determines the conditions for reflection and refraction of sound at the boundary of two media. Acoustic impedance is the general concept of wave resistance, equal to the ratio of sound pressure in a plane traveling wave to the vibrational velocity of the particles of the medium. In the absence of dispersion of the sound velocity, it is independent of the waveform and is expressed by the formula [119]

$$Z = \rho \cdot v, \qquad (2.28)$$

where ρ is the density of the medium; and v is the speed of sound.

Under normal incidence of an elastic wave on the boundary of two media, the reflection coefficients R and transmission coefficient T of wave propagation are determined by the Fresnel formulas

Table 2.2. Acoustic characteristics of materials of absorbers of elastic waves for acousto-optical cells [120]

Name	Symbol	Density ρ $(10^3$ kg/ $m^3)$	Velocity of elastic waves		Wave resistance	
			Longitudinal $V_L(10^3$ m/s)	Shear V_S $(10^3$ m/s)	Longitudinal Z_L $(10^5$ kg/m² s)	Shear Z_S $(10^5$ kg/ m² s)
Aluminum	Al	2.6889	6.26	3.08	169	83.2
Gold	Au	19.32	3.24	1.2	626	232
Indium	In	7.31	2.56	0.81	187	59
Copper	Cu	8.96	4.70	2.26	418	201
Tin	Sn	7.29	3.32	1.67	242	122
Lead	Pb	11.336	2.16	0.70	246	80
Silver	Ag	10.50	3.60	1.59	380	167
Molybdenum	Mo	10.22	5.67	3.51	511	316
Brass		8.2–8.85	4.43	2.12	361	172
Cast iron		7.0–7.2	4.50	2.40	350	187
Epoxy resin	ED-5	1.1	2.58	1.22	29	13
Rubber		0.91	1.48		13	

$$R = (Z_2 - Z_1) / (Z_2 + Z_1), \quad T = 2Z_2/(Z_2 + Z_1), \qquad (2.29)$$

where Z_1, Z_2 are the acoustic impedances (acoustic wave impedances) of these media. Heat sink absorbers are usually made of metals and their alloys. Table 2.2 shows the acoustic characteristics of the absorber materials.

The production of the bevel at the end of the cell and the attachment to it of an acoustic absorber, which simultaneously plays the role of a heat sink, is carried out as follows. The end of the acoustic cell is ground at an angle in the plane of the acousto-optical interaction. The bevel angle should reflect the part of the sound beam that did not pass into the acoustic absorber at such an angle that the Bragg condition of the acousto-optical interaction with the incident optical radiation is not satisfied. The absorber material and the method of its attachment to the cell are selected in such a way that the reflected acoustic beam is as small as possible in intensity.

Usually, various metals or their alloys are selected for the absorber, which have an acoustic impedance close to that of the acousto-optical cell. Using the parameter table, you can select the absorber material. The impedance of the absorber material is selected close to the impedance of the crystal in the direction of propagation of the elastic waves. It should be taken into account that the absorber material must also have good thermal conductivity for removing heat generated by absorbing acoustic vibrations. It must also be taken into account that when the elastic wave falls at an angle, it can be converted and have shear and longitudinal components when reflected.

2.3.8. Sputtering of the upper electrode

Sputtering of the upper electrode of the piezoelectric transducer is necessary to create an acoustic beam of a certain size and shape in the acousto-optical element. For example, the rhomboid shape of the top electrode of the piezotransducer makes it possible to suppress side lobes in the far zone, as is the case with antennas [121].

Sputtering of the upper electrode is carried out in a vacuum through a mask, which is usually cut from an aluminium foil by a laser or a die.

Calculation of the thickness of the connecting layers of the piezoelectric transducer and the crystalline element can be carried out in a manner analogous to the calculation of antireflection coatings for optical radiation. When welding, the thickness of the bonding metal layer is usually about 1 μm, therefore, considering the speed of sound in the bonding layer, half-wave matching thicknesses must be taken into account only in acousto-optical devices operating at frequencies above 500 MHz. The half-wave thickness of indium at a frequency of 500 MHz for shear oscillations will be 0.81 μm, and for longitudinal oscillations it will be 2.56 μm (see Table 2.1). At frequencies below 500 MHz, it is desirable to obtain binder layers of not more than 1–2 μm. This is due to the high attenuation of elastic waves in indium, as the main binding metal.

2.4. Acousto-optical modulator for a pulsed television image formation system

An acousto-optica spatial light modulator for a pulse imaging system of a line must meet a number of requirements. First, the length of the modulator's sound line must satisfy the relation (1.3). This requirement, on the one hand, restricts the choice of the material of the sound transmission line by a small propagation velocity of ultrasound, on the other hand, the attenuation of ultrasonic waves at a wavelength L in the selected frequency range should not be large. Secondly, the modulator should have a sufficiently high efficiency of controlling the light beam, which implies choosing a material of a sound transmission line with high coefficients of acousto-optical quality M_1 and M_2 [93], as well as small losses of light for absorbing the working wavelength. To reduce the loss of light by reflection, it is necessary to be able to apply interference antireflection coatings to optical modulator windows that withstand a high power density of laser radiation. Thirdly, the AOM sound line material should not have significant phase inhomogeneities leading to distortion of the wave front of the light wave carrying information.

Analysis of the currently existing media for acousto-optical interaction shows that the paratellurite crystal (TeO_2) satisfies the highest requirements in the ultrasonic frequency range to 100–150 MHz. This optically uniaxial crystal transparent in the visible and near infrared wavelength ranges has a number of unique acousto-optical properties [122, 123]. Thus, for example, the velocity of a slow shear ultrasonic wave along the crystallographic direction [110] is 0.616×10^3 m/s, which allows a modulated ultrasound signal of 52 µs duration along this direction for a length of 32 mm.

The use of a slow shear ultrasonic wave for the diffraction of light by sound is possible with an anisotropic geometry of the acousto-optical interaction. For anisotropic scattering geometry, in contrast to isotropic, the wave vectors of the incident and diffracted light waves lie on different surfaces of the wave vectors of the crystal. Diffraction of light occurs with rotation of the plane of polarization. The optical activity of the TeO_2 crystal leads to the fact that when light propagates near the optical axis, normal waves are waves with elliptical polarization. As a result of anisotropic scattering, the ellipticity, the direction of rotation and the azimuth of the ellipse of polarization of light waves change. In more detail, the polarization of interacting light waves in a TeO_2 crystal will be discussed in Chapter 3, here we emphasize that to effectively use the light power for acousto-optical interaction in a given crystal, it is necessary to match the polarization of the light incident on the crystal with the scattering required for the chosen scattering geometry.

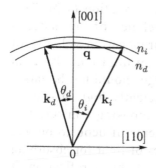

Fig. 2.17. Vector diagram for anisotropic geometry of light scattering in a TeO$_2$ crystal.

The light scattering diagram in an anisotropic TeO$_2$ crystal can be represented in the form of a vector triangle (Fig. 2.17).

In this diagram, θ_i and θ_d are the angles of incidence and diffraction of light waves inside the crystal, measured from the crystallographic direction **[001]**, which coincides with its optical axis; n_i and n_d are the refractive indices for the incident and diffracted waves, respectively.

From expressions (2.1), one can obtain conditions for the effective Bragg diffraction of light by sound, at which the maximum intensity of light in the diffracted beam is ensured. These conditions are determined by the joint solution of the following equations:

$$n_d \cdot \cos\theta_d - n_i \cdot \cos\theta_i = 0,$$

$$n_d \cdot \sin\theta_d + n_i \cdot \sin\theta_i = \frac{\lambda \cdot f}{v}, \qquad (2.30)$$

$$n_i^2 = \left(\frac{\sin^2\theta_i}{n_e^2} + \frac{\cos^2\theta_i}{n_o^2 \cdot (1+\delta)^2} \right)^{-1},$$

$$n_d^2 = \left(\frac{\sin^2\theta_d}{n_o^2} + \frac{\cos^2\theta_d}{n_o^2 \cdot (1-\delta)^2} \right)^{-1}, \qquad (2.31)$$

$$v = \left[\frac{(c_{11} - c_{12})}{2} \cdot \frac{1}{\rho} \right]^{1/2}, \quad \delta = \frac{\Delta n}{2 \cdot n_o},$$

where n_o and n_e are the refractive indices of ordinary and extraordinary light waves in a crystal; Δn is the change in the refractive indices caused by gyrotropy for the light waves propagating along the optical axis of the TeO$_2$ crystal; λ is the wavelength of light in vacuum; c_{11}, c_{12} are the elastic constants of the TeO$_2$ crystal; ρ is the density of the sound transmission material; f is the frequency of acoustic waves, v is the speed of sound in the material of the sound line.

When the gyrotropic crystal is diffracted near the optical axis, in view of the smallness of Δn and the angles θ_i, θ_d, the expressions for the refractive indices n_i and n_d can be simplified:

$$n_{i,d} = n_o \cdot \left[1 \pm \sqrt{a^2 + \tilde{c}^2 \cdot \theta_{i,d}^4} + \tilde{c} \cdot \theta_{i,d}^2 \right], \qquad (2.32)$$

where

$$a = \frac{G_{33} \cdot n_o^2}{2}, \quad \tilde{c} = \frac{(n_e^2 - n_o^2)}{4 n_e^2}, \qquad (2.33)$$

G_{33} is the component of the pseudotensor of the gyration in the [001] direction of the TeO_2 crystal. Setting the carrier frequency of sound f and solving the system of non-linear equations (2.30) ÷ (2.32), one can find the angles θ_i and θ_d at which anisotropic diffraction occurs. For example, Fig. 2.18 shows the results of calculating such an angular-frequency dependence for a TeO_2 crystal at $\lambda = 510.6$ nm, $n_e = 2.476$, $n_o = 2.314$, $G_{33} = 3.745 \cdot 10^{-5}$.

Choosing angles θ_i as the angles of incidence in the flat portion of the dependence $\theta_i (f)$, we work in the so-called broadband diffraction mode, in which the Bragg diffraction conditions are performed in a wide frequency band Δf for a relatively small range of angles of incidence: $\Delta \theta_i \ll \Delta \theta_d$. When the Bragg conditions are satisfied, the efficiency of the diffraction of light into the +1st diffraction order is determined by the expression

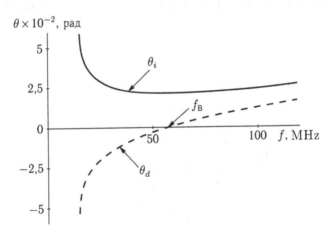

Fig. 2.18. Angular-frequency dependence for anisotropic Bragg diffraction of light on sound in a TeO_2 crystal near its optical axis.

$$I = \frac{I_{+1}}{I_0 + I_{+1}} \cong \sin^2(B \cdot \sqrt{M_2 \cdot P_a}), \qquad (2.34)$$

where I_{+1} is the intensity of light diffracted by +1st order at frequency f; I_0 is the intensity of non-diffracted light at the AOM output, B is a constant depending on the wavelength of light and the size of the sound beam, $M_2 = \dfrac{n_o^6 \cdot p_{eff}^2}{\rho \cdot v^3}$ is a constant value that characterizes the acousto-optical quality or quality factor of the sound transmission material, P_a is the acoustic power. The frequency f_B at which the conditions $\theta_d = 0$, $\dfrac{d\theta_i}{df} = 0$ are satisfied is called the frequency of the two-phonon interaction or the degeneracy frequency. From the expressions (2.30), (2.32), we can find that

$$f_B' = \frac{v}{\lambda_0} n_o \sqrt{(n_o^2 + n_e^2) \cdot G_{33}}. \qquad (2.35)$$

For the graph shown in Fig. 2.18 $f_B' = 57.9$ MHz.

At this frequency, conditions are satisfied for effective diffraction of light by sound into the second diffraction order.

The geometry of the acousto-optical interaction, which takes place in a real device for AOM from TeO_2, is shown in Fig. 2.19. In this case, along the [110] direction of the TeO_2 crystal, coinciding with the X axis of the optical system, a slow ultrasound wave propagates along its [$\bar{1}$10] axis and coincides with the Y axis.

A light beam is incident on the modulator parallel to the scattering plane ($\bar{1}$10). The angle between the wave vector **k** and its projection

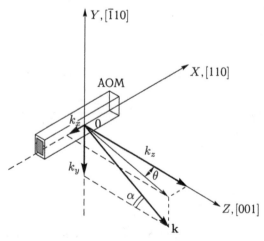

Fig. 2.19. Geometry of the acousto-optical interaction in a TeO_2 crystal.

onto the ($\bar{1}10$) plane is α, and the angle between the same projection and the Z axis is θ.

The TeO$_2$ crystal is distinguished by a high degree of anisotropy of the elastic and optical properties over various crystal physical directions. The choice of the optimum directions of propagation of elastic and light waves from the point of view of the diffraction efficiency was made at aid to computer calculations [124]. Calculations show that for the scattering plane coinciding with the ($\bar{1}10$) plane of the TeO$_2$ crystal, for the anisotropic diffraction of light by a slow shear wave propagating in the [110] direction and for the directions of the wave vector of light near the optical axis, the reduced acousto-optical quality factors M_1 and M_2 are equal to 68 and 793, respectively (for a wavelength of light of 632.8 nm) [123]. This great efficiency allows us to use small ultrasound powers (1 W or less) to control acousto-optical devices. On the other hand, the high anisotropy of the elastic properties of the TeO$_2$ crystal leads to the need for strict orientation of the faces of the sound line. Thus, the misorientation of the face on which the piezoelectric transducer excites a slow shear wave by 0.5° relative to the [110] axis in the (001) plane of the TeO$_2$ crystal leads to a deflection of the energy flow of the elastic wave in the final device by 25° in the indicated plane, which is the plane optical aperture of the AOM. The latter leads to the fact that for small ratios h/L, where h is the height of the piezoelectric transducer, L is the length of the acoustic line, the acousto-optical interaction will be realized only with part of the light flux incident on the modulator. Similar results can be obtained by the wedge shape of the intermediate bonding layers between the sound pipe and the piezoelectric transducer. In this connection, the construction of an acousto-optical spatial modulator on a TeO$_2$ crystal for a long crystal length, as well as the technology of its fabrication, requires improvement.

The attenuation of a slow sound wave at a length of a 35 mm sound line at a frequency of 80 MHz is approximately $6 \div 6.5$ dB. Despite the fact that attenuation for this type of ultrasonic wave is sufficiently large (290 dB/cm $\cdot$ GHz^2), the natural gyrotropy of the crystal makes it possible to efficiently use the broadband anisotropic scattering geometry at frequencies of 100 MHz and below.

The technology of high quality TeO$_2$ crystal growth is now well developed in a number of laboratories in the world. The length of the growing crystal boule along the crystallographic direction [110] can reach $60 \div 100$ mm with a diameter of 50 mm [125]. Scattering inside the crystal is practically not observed. Other types of scattering associated with the processing of the surface of optical windows or with parasitic reflections within a crystal can be minimized. The quality of antireflection coatings achieved at present allows to achieve a transmittance of AOM from TeO_2 of 99.5% for the visible wavelength range.

Studies on the optical stability of TeO_2 crystals show that crystals survive a periodic operation at a frequency of $1 \div 3$ Hz at laser radiation power densities of up to 1 GW/cm^2 [126]. Improvement of the technology of manufacturing and construction of modulators from TeO_2 [90] made it possible to achieve high resistance to external influences in the form of mechanical vibrations, impacts on the body, accelerations, etc. With proper operation, the lifetime of the AOM expressed in years.

Structurally, the acousto-optical modulator contains a sound line made of a paratellurite crystal, in which two facets are polished for light and illuminated for laser wavelengths. A piezoelectric transducer made of lithium niobate $LiNbO_3$ X is attached to the end of the crystalline cell by means of thermal compression welding, a cut that excites elastic vibrations, and to the opposite one − an absorber of elastic waves from aluminum. The thickness of the piezoelectric transducer is ground to 25 µm. Electrodes from aluminium are sprayed on it. The shape of the upper electrode of the piezoelectric transducer is made in the form of an elongated hexagon for suppressing the lateral acoustic lobes. The cell is placed in the holder, which also contains a device that matches the output of the generator to the piezotransducer, and a coaxial connector for supplying an electrical signal.

Figure 2.20 shows the AOM sound tube from the TeO_2 crystal with a welded piezotransducer and sputtered electrodes. Figure 2.21 presents a photo of an acousto-optical modulator for a laser television projector mounted in the housing.

2.5. Devices for the scanning of the light beam

The selection of the device for deflecting the light beam according to the frame depends on the requirements imposed on the entire information display system. The parameters that determine the operation of such a device include the number of solvable elements, the linearity of the deviation, the efficiency, the possibility of combining the image at different wavelengths. The number of

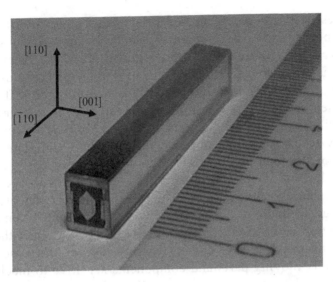

Fig. 2.20. Photograph of the AOM sound duct from a TeO$_2$ crystal with a welded piezotransducer and coated electrodes.

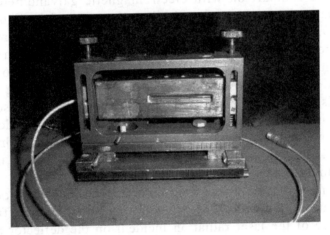

Fig. 2.21. Acousto-optical modulator for laser television installation in the housing.

decidable elements of any deflector during scanning is determined by the expression

$$N_{def} = \frac{\Delta\varphi_{sc}}{\Delta\Psi_0}, \tag{2.36}$$

where $\Delta\varphi_{sc}$ is the scanning angle, $\Delta\Psi_0$ is the divergence of the light beam at the output from the deflector.

For electromechanical deflectors (mirror galvanometers, rotating polyhedral mirror drums), the scanning angles can reach several tens of degrees. With the divergence of a light beam close to diffraction, the number of solvable elements in a frame can reach several thousand. If the required resolution is not so great, as, for example, in a television standard with 625 lines in a frame, one can use light beams with a divergence significantly exceeding the diffraction limit. When using mirror electromagnetic galvanometers, in addition to providing high linearity and stability of the forward sweep, it is necessary to ensure a fast mirror return to the initial state during the time of the blanking pulse (1.6 ms or less). Currently, there is a large selection of electromagnetic galvanometers with acceptable characteristics for the deflection of laser radiation characteristics (Cambridge Technology Inc., General Scanning Inc., GSI Lumonics).

The use of acousto-optical deflectors (AOD) to deflect the light beam along the frame is convenient because of the absence of mechanically moving parts and components sensitive to mechanical vibrations and pressure drops, as well as the possibility of rapid electronic control of the scanning parameters. A great advantage of AOD in comparison with electromagnetic galvanometers is a significantly shorter time of the backward motion of the light beam. For AOD, this time is determined by the time of the sound wave travel through the length of the sound line and is equal to just a few tens of microseconds. The angle of scanning of the acousto-optical deflector does not exceed several degrees, so when working with an AOD it is necessary to aim at minimum divergence of the laser beam and take into account all the mechanisms that lead to a broadening of the laser radiation pattern along the Y coordinate. A certain drawback of AOD is the thickness of the sound transmission, which must be taken into account when calculating the optical system.

For a fixed angle $\Delta\varphi_{sc}$ the maximum number of solvable elements at the output from the AOD is obtained with the diffraction divergence of the laser radiation incident on the deflector and the homogeneous distribution of the diffracted field on the rectangular output aperture:

$$N_k = \frac{D \cdot \Delta\varphi_{sc}}{\lambda_0}, \tag{2.37}$$

where λ_0 is the wavelength of light in vacuum, D is the aperture of the deflector in the scattering plane. If the field distribution in the output beam of the deflector is non-uniform, for example, due to the decrease in the diffraction efficiency due to damping of the elastic wave, then, as shown, for example in Ref. [127], the limiting resolution decreases.

The value of the scanning angle can be determined by solving a system of equations expressing the law of conservation of energy and momentum for interacting light and sound waves in a crystal (2.3). For geometric interpretation of this solution, it is convenient to use vector scattering diagrams for the selected scattering plane. Each vector diagram corresponds to the limiting case of the diffraction theory in which the Bragg conditions for incident and diffracted light waves are satisfied. From the vector scattering diagram, the relations between the amplitudes of the wave vectors of the incident $k_i = \dfrac{2\pi}{\lambda_0} n_i$ and diffracted $k_d = \dfrac{2\pi}{\lambda_0} n_d$ light in the crystal, the incidence angles θ_i and the diffraction angles θ_d, and also the wave vector of the elastic wave $\mathbf{q} = \dfrac{2\pi \cdot f}{v} \cdot \mathbf{r}$, where $\mathbf{r}$ is the direction of its propagation. The change in one of the quantities entering the vector triangle entails a change in the others.

For the AOD it is usually necessary to use a mode of operation in which the direction and magnitude of the incident wave vector k_i remain constant and the direction of the diffracted light beam is achieved by changing the frequency and direction of propagation of the ultrasonic wave. The required change in the direction of propagation of an ultrasonic wave is most simply achieved at low frequencies due to the diffraction divergence of the sound beam. Normally, an ultrasonic signal whose frequency varies linearly (a chirp signal) is applied to the light deviation from the frame to the AOD.

Many studies have been devoted to investigating the features of the geometry of the acousto-optical interaction in crystals of various classes, in particular [110, 128]. Deflectors using anisotropic diffraction of light have a definite advantage over deflectors with isotropic diffraction. This advantage is expressed by the fact that when operating near the frequency f'_B, the required range of deflection angles can be achieved with a lower ultrasonic divergence, and, consequently, with lower power inputs. Of the anisotropic deflectors, many authors [129, 130] single out a deflector on a TeO_2 crystal, for which the frequency $f'_B = f_B$ is small (57.9 MHz at $\lambda = 510.6$ nm). Broadband scattering geometry can be implemented at fairly low ultrasonic frequencies in the 50 to 100 MHz band with a high diffraction efficiency (70 ÷ 80%). As a rule, the range of controlled audio frequencies does not exceed one octave in order to eliminate the second diffraction order in the scanning region.

When operating in a wide frequency band in an anisotropic deflector from TeO_2, it is possible to transfer the energy of the diffracted light to the second diffraction order. This is most pronounced at high diffraction efficiency at ultrasonic frequencies near f_B, which leads to a dip on the amplitude–frequency characteristic of the deflector. To eliminate this effect, it was suggested in [131] to use non-axial sections of the TeO_2 crystal in which the scattering plane is rotated through an angle $\phi \cong 6°$ with respect to the $(\bar{1}10)$ plane. In this case, the frequency of the two-phonon interaction turns out to be outside the working frequency band, and there is no dip in the amplitude-frequency characteristic of the deflector. In such a device [132], the diffraction efficiency was 92% with a control bandwidth of 50 MHz and an electric power input of 210 mW.

The limiting number of solvable elements in this deflector can be determined from the relation

$$N_k = \frac{\Delta f \cdot D}{v},$$
(2.38)

where D is the size of the optical aperture of the deflector along the Y coordinate. For a deflector with parameters $D = 15$ mm, $\Delta f = 50$ MHz and $v = 0.65 \cdot 10^3$ m/s, we obtain $N_k \cong 1154$ by the Rayleigh criterion, which is quite applicable for information display systems operating in the high definition standard.

For effective operation of the AOD, certain requirements are imposed on the parameters of the light beam. So, in the plane of deflection along the frame, the beam of light incident on the deflector must be parallel and have dimensions corresponding to the dimensions of the sound field in the AOD crystal, and the polarization of the incident light must correspond to the acousto-optical interaction required for the chosen geometry. For a TeO_2 deflector with a 6° cut, the polarization of the incident light should be close to linear with the polarization axis lying in the plane of the deflector scattering (plane $(\bar{1}10)$). If, after the AOD, the diffracted light has a polarization close to circular, then to effectively deflect light through the frame with the help of an AOD of this type, it must be converted into linear by means of, for example, a quarter of the wave plate.

Of great importance in the design of the AOD is the uniformity and width of the sound field within the working aperture of the deflector. For undistorted reproduction of the image on the screen, the deflector must equally effectively deflect the light beams related to the diffraction components from which this image is formed. The working aperture of the AOD, determined by the width of the sound column in the deflector, is a spatial frequency filter and determines the bandwidth of the TV image forming system.

2.6. Variants of construction of an optical system for an acousto-optical device for the formation of a television image with a pulsed laser

The optical scheme of an acousto-optical television information display device based on a pulsed laser should provide the following functions:

1) matching the sizes of the laser beam in two coordinates with the dimensions of the ultrasonic beam in the acousto-optical modulator;
2) the projection of the image of the line from the output plane AOM to the screen along the X and Y coordinates with magnification factors M_x and M_y;
3) matching the sizes of the light beam at the output of the AOM with the dimensions of the deflecting element in the case of an electromechanical system of deflection along the frame or with the dimensions of the sound field in the acousto-optical deflector;
4) matching the polarization of the light beam incident on the modulator and deflector, with the polarization required for effective acousto-optical interaction;
5) coordination of the two dimensions of the TV screen on the screen in accordance with the adopted standard.

Figure 2.22 shows one of the variants of the optical scheme of the device for displaying TV information with the configuration of the sizes of the light beam in two mutually perpendicular planes is presented.

In this scheme, radiation from a pulsed laser is fed to a telescopic system consisting of lenses L_1 and L_2, after which the diameter of the light beam corresponds to the length of the acousto-optical modulator. The cylindrical lens L_3 creates a constriction of the light beam near which the modulator (AOM) is placed. With this

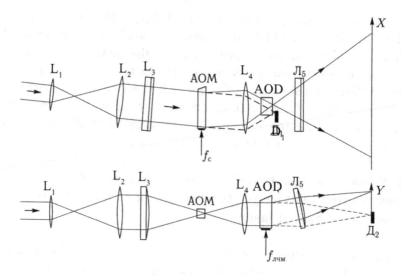

Fig. 2.22. A possible variant of the optical scheme of the device for displaying TV information with the configuration of the sizes of the light beam in two mutually perpendicular planes.

position of the modulator, it is sufficient to form an acoustic beam having small transverse dimensions for efficient control of the laser radiation. At the same time, such a configuration allows selecting a region on the modulator aperture with small phase distortions and a uniform distribution of the sound field. The output aperture of the modulator is the object plane of the L_4 lens, which forms the image of the line on the screen. The choice of the focal length of the L_4 lens approximately coinciding with the distance between A_4 and AOM makes it possible to form a parallel beam of light along one coordinate, which falls on the acousto-optical deflector (AOD). In this case, along another coordinate, the deflector is placed near the Fourier plane lenses of L_4. With this arrangement of the deflector, its transverse dimensions are minimal. The transverse dimensions of the sound column in the deflector should be such as to reject the entire spectrum of spatial frequencies from the modulated light signal, which is formed by the L_4 lens. The deflector acts as a spatial frequency filter, which rejects only the spectral diffraction orders necessary for imaging. The unused light orders are absorbed by the diaphragm D_1. The cylindrical lens L_5 serves to focus the light beam on one coordinate to the screen, and also to match the angular dimensions of the television raster in accordance with the standard.

In addition to the listed elements, this scheme can contain optical components that change the polarization of the light beam, as well as the diaphragm, rotary mirrors, prisms, etc., and it can vary depending on the size of the image being formed, the distance to the screen and the shape of the raster.

This scheme can be taken as a basis for theoretical consideration of the process of image formation using a pulsed laser. Theoretical consideration of the process of image formation in a system with an acousto-optical modulator from TeO_2 and a pulsed laser is necessary for solving a number of practical problems, including:

1) calculation of the frequency–contrast characteristic of the system;

2) finding the distribution of light intensity for one line in the plane of the screen and estimating the number of solvable elements of the system by frame;

3) determination of the influence on the frequency contrast characteristic and the number of solvable elements of the system of such parameters of the light source as the pulse duration and radiation polarization;

4) determination of the dependence of the efficiency of diffraction and the number of solvable elements of the system on the angles of incidence of light on the modulator in the plane of deflection of light along the frame;

5) estimation of the influence of the non-linearity of the diffraction process of light in AOM from TeO_2 on the frequency contrast characteristic of the system and the determination of the nature of the nonlinear distortions arising in the high amplitude of the ultrasonic signal in the image being formed;

6) the solution of this problem is complicated by the fact that to describe the process of diffraction of light by sound in our case it is impossible to confine ourselves to the case of low diffraction efficiency or only to the Bragg conditions of acousto-optical interaction. It is necessary to consider the so-called intermediate diffraction case and, in addition, take into account the anisotropic and gyrotropic properties of the interaction medium.

Theoretical consideration of the process of image formation in the acousto-optical system with the pulsed coherent light source

3.1. Diffraction of a plane light wave by ultrasound in a gyrotropic TeO$_2$ crystal

The problem of the diffraction of light by ultrasound was considered by many authors [133–143]. At present, there are two most common approaches to solving this problem, which, in principle, allow one to obtain a solution with a predetermined degree of accuracy. The first approach is based on solving a system of coupled differential-difference equations describing the interaction between different diffraction orders. This method found the most complete expression in [144–148], and also in [149–151]. Another approach is based on solving the integral equation for a field obtained by introducing equivalent currents and expanding the desired field over plane waves [152–155], which allows one to obtain an analytical expression for the diffracted field in the form of relatively rapidly converging series. The practical interest that causes the TeO$_2$ crystal requires consideration of the problem of diffraction of light by ultrasound in an anisotropic medium with gyrotropic properties. Usually when considering the diffraction of light in such a crystal, the authors confine themselves to an investigation of the Bragg regime. In this chapter we give a generalized solution of the diffraction problem in the case of an anisotropic medium with gyrotropy, based on the method proposed in Refs. [152–155], which ultimately yields an expression for the diffracted field after an acousto-optical modulator from TeO$_2$ with its amplitude modulation with respect to the harmonic law in the intermediate diffraction regime.

3.1.1. Statement of the problem and derivation of the integral equation for the field

Let us consider a homogeneous infinite anisotropic non-magnetic and non-conducting medium with a weak spatial dispersion. The material equation for such a medium has the form [156]:

$$\mathbf{D} = (\widehat{\varepsilon^0} + i\widehat{\gamma}\cdot\mathbf{k})\mathbf{E}; \quad D_i = (\varepsilon_{ij}^0 + i\gamma_{ijl}\cdot\kappa_l)E_j, \qquad (3.1)$$

where $\mathbf{D}$ is the displacement vector, $\mathbf{E}$ is the electric field vector, $\widehat{\varepsilon^0}$ is the permittivity tensor of the medium, $\widehat{\gamma}$ is the third-rank tensor, antisymmetric in the first two indices, characterizing the gyrotropy of the medium, and $\mathbf{k}$ is the wave vector of the electromagnetic field.

We define in the given medium a coordinate system $x_1 = X$, $x_2 = Y$, $x_3 = Z$, connected with the crystal-physical system x_1^0, x_2^0, x_3^0 by the transformation matrix α_{ij}: $x_i = \alpha_{ij}x_j^0$. In the general case, in the plane $X0Z$ of this medium, which we will call the scattering plane, two elliptically polarized plane light waves can propagate in the direction $\mathbf{m}$, which makes an angle θ with the Z axis [157]

$$\mathbf{E}_{k_d} = \mathbf{U}_{k_d}\cdot e^{i(\mathbf{k}_d\cdot\mathbf{r} - \omega t)}, \qquad (3.2)$$

where

$$\mathbf{k}_d\cdot\mathbf{r} = \frac{2\pi}{\lambda_0}n_d\cdot(\cos\theta\cdot z - \sin\theta\cdot x), \quad d=1, 2;$$

$\mathbf{r}$ is the radius vector, λ_0 is the wavelength of light in vacuum; ω is the circular frequency of the light wave; $\mathbf{U}_{k_d}$ is the vector of the amplitude of the light wave; n_d are the refractive indices of the medium for two elliptically polarized waves propagating in the direction $\mathbf{m}$.

We will consider in this medium a volume V bounded by the planar surfaces $S_1\left(z = -\dfrac{L}{2}\right)$ and $S_2\left(z = -\dfrac{L}{2}\right)$, which have infinite dimensions in X and Y (see Fig. 3.1).

In this medium, for $|z| \leqslant \dfrac{L}{2}$ an ultrasonic wave propagates in the direction $\mathbf{r}'$

$$\mathbf{u}(\mathbf{r},\ t) = A(\mathbf{r}' - \mathbf{v} \cdot t) \cdot \mathbf{u}^0 \cdot e^{i(\mathbf{q} \cdot \mathbf{r} - \Omega \cdot t)}, \tag{3.3}$$

where $\mathbf{u}(\mathbf{r},\ t)$ is the displacement vector, $A\,(\mathbf{r}' - \mathbf{v} \cdot t)$ is the amplitude distribution moving in the crystal with a velocity $\mathbf{v}$ whose law of variation is determined by external control, $\mathbf{u}^0$ is the polarization vector characterizing the direction of the local displacements of particles environment, $\mathbf{q}$ is the wave vector of the elastic wave, $\Omega = 2\pi \cdot f$ is the circular frequency. This wave causes a deformation, which in turn leads to a change in the components of the dielectric medium tensor by an amount

$$\Delta \eta_{ij} = p_{ijfp} \cdot \xi_{fp} = \xi \cdot p_{ijfp} \cdot \xi_{fp}^0, \tag{3.4}$$

where

$$\xi = A(\mathbf{r}' - \mathbf{v} \cdot t) \cdot e^{i(\mathbf{q} \cdot \mathbf{r} - \Omega \cdot t)}, \tag{3.5}$$

$$\xi_{fp}^0 = \frac{1}{2}\left(\frac{\partial u_f^0}{\partial x_p} + \frac{\partial u_p^0}{\partial x_f} \right) = i \cdot q_f \cdot u_p^0, \tag{3.6}$$

p_{ijfp} and ξ_{fp} are the components of the photoelasticity tensors $\widehat{p}$ and deformation $\widehat{\xi}$ in the chosen coordinate system. Neglecting the

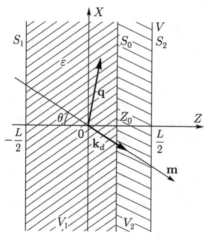

Fig. 3.1. Geometry of the interaction of light with sound.

acoustical non-linearity of the interaction, under the condition $\varepsilon_{ik} \cdot \Delta\eta_{kj} \ll 1$ and $\gamma_{ijl} \ll 1$, we will consider in this medium the volume V bounded by the perturbed tensor of the permittivity $\widehat{\varepsilon'}$ of the medium can be written:

$$\widehat{\varepsilon'} = \widehat{\varepsilon} - \widehat{\varepsilon} \cdot \widehat{\Delta\eta} \cdot \widehat{\varepsilon} \cong \widehat{\varepsilon} - \widehat{\varepsilon^0} \cdot \widehat{\Delta\eta} \cdot \widehat{\varepsilon^0}, \tag{3.7}$$

where $\widehat{\varepsilon} = \widehat{\varepsilon^0} + i\widehat{\gamma} \cdot \mathbf{k}$, $\widehat{\varepsilon^0}$ is the unperturbed tensor of the permittivity of the medium in the absence of gyrotropy.

We write down the Maxwell equations for the electromagnetic field in an anisotropic non-magnetic and non-conducting medium with gyrotropy in the presence of a perturbation $\widehat{\Delta\eta}$ under the following assumptions: $\Omega \ll \omega$, $\varepsilon_{ik} \cdot \Delta\eta_{kj} \ll 1$, $\gamma_{ijl} \ll 1$. Then we have the following equations:

$$\begin{cases} \text{rot } \mathbf{E} = \dfrac{i\omega}{c}\mathbf{H}, \\[2mm] \text{rot } \mathbf{H} = \dfrac{i\omega}{c}\widehat{\varepsilon}\cdot\mathbf{E} + \dfrac{i\omega}{c}\widehat{\varepsilon^0}\cdot\widehat{\Delta\eta}\cdot\widehat{\varepsilon^0}\cdot\mathbf{E}, \end{cases} \tag{3.8}$$

where c is the speed of light in vacuum.

We introduce equivalent currents

$$\frac{4\pi}{c}\mathbf{j}_{eq} = \begin{cases} \dfrac{i\omega}{c}\widehat{\varepsilon^0}\cdot\widehat{\Delta\eta}\cdot\widehat{\varepsilon^0}\cdot\mathbf{E}, & |z| \leqslant \dfrac{L}{2}, \\[3mm] 0, & |z| > \dfrac{L}{2}, \end{cases} \tag{3.9}$$

and reduce the solution of the Maxwell equations to the solution of an integral equation of the form:

$$\mathbf{E} = \mathbf{E}_i + \mathcal{L}(\mathbf{j}_{eq}) = \mathbf{E}_i + \mathcal{L}\left(\frac{i\omega}{c}\widehat{\varepsilon^0}\cdot\widehat{\Delta\eta}\cdot\widehat{\varepsilon^0}\cdot\mathbf{E}\right), \tag{3.10}$$

where $\mathbf{E}_i$ is the incident light wave, $i = 1, 2$; $\mathcal{L}$ is an integral operator. The derivation of the integral equation for the diffracted field is based on the application of the Lorentz lemma to the volumes V_1 and V_2 (see Fig. 3.1).

To derive the Lorentz lemma for an anisotropic gyrotropic medium, we consider two fields $\mathbf{E}_1 = \mathbf{U}_1 \cdot e^{i(\mathbf{k}_1 \cdot \mathbf{r} - \omega t)}$, $\mathbf{H}_1 = \mathbf{G}_1 \cdot e^{i(\mathbf{k}_1 \cdot \mathbf{r} - \omega t)}$

and $E_2^* = U_2^* \cdot e^{-i(k_2 \cdot r - \omega t)}$, $H_2^* = G_2^* \cdot e^{-i(k_2 \cdot r - \omega t)}$, which are excited by the field sources j_1 and j_2^* in a medium with the permittivity tensor $\hat{\varepsilon} = \hat{\varepsilon^0} + i\hat{\gamma} \cdot k$ and satisfy the Maxwell equations:

$$\text{rot } E_1 = \frac{i\omega}{c} \cdot H_1, \quad \text{rot } H_1 = -\frac{i\omega}{c} \cdot \hat{\varepsilon} \cdot E_1 + \frac{4\pi}{c} \cdot j_1; \tag{3.11}$$

$$\text{rot } E_2^* = -\frac{i\omega}{c} \cdot H_2^*, \quad \text{rot } H_2^* = \frac{i\omega}{c} \cdot \hat{\varepsilon} * \cdot E_2^* + \frac{4\pi}{c} \cdot j_2^*, \tag{3.12}$$

where (*) stands for complex conjugation.

We multiply the first equation (3.11) scalarly by H_2^*, and the second equation (3.12) by $-E_1$ and add. We get:

$$H_2^* \cdot \text{rot } E_1 - E_1 \cdot \text{rot } H_2^* = \frac{i\omega}{c}(H_1 \cdot H_2^* - E_1 \cdot \hat{\varepsilon} * \cdot E_2^*) - \frac{4\pi}{c} j_2^* \cdot E_1. \tag{3.13}$$

Similarly, we multiply the first equation (3.12) by H_1, and the second equation (3.11) by $-E_2^*$. After addition, we get:

$$H_1 \cdot \text{rot } E_2^* - E_2^* \cdot \text{rot } H_1 = \frac{i\omega}{c}(E_2^* \cdot \hat{\varepsilon} \cdot E_1 - H_1 \cdot H_2^*) - \frac{4\pi}{c} j_1 \cdot E_2^*. \tag{3.14}$$

Adding (3.13) and (3.14) and using the equality

$$\text{div}(A \times B) = B \cdot \text{rot } A - A \cdot \text{rot } B, \tag{3.15}$$

we obtain the relation

$$\text{div}(E_1 \times H_2^*) + \text{div}(E_2^* \times H_1) =$$
$$= \frac{i\omega}{c}(E_2^* \cdot \hat{\varepsilon} \cdot E_1 - E_1 \cdot \hat{\varepsilon} * \cdot E_2^*) - \frac{4\pi}{c}(j_2^* \cdot E_1 + j_1 \cdot E_2^*). \tag{3.16}$$

Consider the expression

$$E_2^* \cdot \hat{\varepsilon} \cdot E_1 - E_1 \cdot \hat{\varepsilon} * \cdot E_2^* =$$
$$= E_2^*(\hat{\varepsilon^0} + i\hat{\gamma} \cdot k)E_1 - E_1(\hat{\varepsilon^0} - i\hat{\gamma} \cdot k)E_2^* = U \cdot e^{i(k_1 \cdot r - k_2 \cdot r)}, \tag{3.17}$$

where

$$U = U_{2i}^* \cdot (\varepsilon_{ij}^0 + i\gamma_{ijl} \cdot k_l)U_{1j} - U_{1i} \cdot (\varepsilon_{ij}^0 + i\gamma_{ijl} \cdot k_l)U_{2j}^*. \tag{3.18}$$

Taking into account that the tensor $\widehat{\varepsilon^0}$ is symmetric and the tensor $\hat{\gamma}$ is antisymmetric with respect to the first two indices ij, and also changing the indices $i \rightarrow j, j \rightarrow i$ in the second term (3.18), we obtain $U = 0$. Integrating (3.16) by the volume V occupied by the sources $\mathbf{j}_1$ and $\mathbf{j}_2^*$, and using the Gauss–Ostrogradsky theorem, we obtain the expression:

$$\oint_S (\mathbf{E}_1 \times \mathbf{H}_2^* + \mathbf{E}_2^* \times \mathbf{H}_1) \cdot \mathbf{n} \, dS = -\frac{4\pi}{c} \int_V (\mathbf{j}_2^* \cdot \mathbf{E}_1 + \mathbf{j}_1 \cdot \mathbf{E}_2^*) dV, \tag{3.19}$$

where V is the volume occupied by equivalent currents, S is the surface bounding this volume; $\mathbf{n}$ is the outer normal to the surface S; $\mathbf{E}_{1,2}$; $\mathbf{H}_{1,2}$ are the fields satisfying Maxwell's equations with external currents $\mathbf{j}_{1,2}$, (*) is the sign of complex conjugation.

Expression (3.19) is one of the forms of the Lorentz lemma, which holds for an anisotropic medium with gyrotropic properties.

To find the integral operator $\mathcal{L}$, we use, similarly to [152, 153], the technique for solving the waveguide excitation problem by a layer of equivalent currents $\mathbf{j}_{eq}$ [158]. We find the field in a certain section $S = S_0$ for $z = z_0$ of the volume V occupied by the sources of the field $\mathbf{j}_{eq}$ (see Fig. 3.1). Let us single out two sections near $z = z_0$: $z_0 - \Delta z$ and $z_0 + \Delta z$, where Δz is a small quantity, and remove the field sources from this layer. The entire volume V is then split into two volumes V_1 and V_2, bounded by the planes $S_1 \left(z = -\dfrac{L}{2} \right)$ and $S(z = z_0 - \Delta z)$, $S (z = z_0 + \Delta z)$ and $S_2 \left(z = -\dfrac{L}{2} \right)$, as well as S_{side}. Letting S tend to infinity, let us assume that the fields are zero by S_{side}. By the Lorentz lemma (3.19) applied to fields in each of these volumes in the absence of field sources, we have:

$$\lim_{S \to \infty} \int_\Sigma (\mathbf{E}_{k_d} \times \mathbf{H}_{l_d}^* + \mathbf{E}_{l_d}^* \times \mathbf{H}_{k_d})\mathbf{n} \, dS = 0, \tag{3.20}$$

where $\Sigma = S_1 + S_2 + 2S$ is the total side surface. Since the integrals over the surfaces S_1 and S_2 are fixed quantities, and the choice of S

along the z axis is arbitrary, then

$$\lim_{S\to\infty} \int_S (\mathbf{E}_{k_d} \times \mathbf{H}^*_{l_d} + \mathbf{E}^*_{l_d} \times \mathbf{H}_{k_d})\mathbf{n}\, dS = J(z) = J(0) \cdot e^{i(\mathbf{k}_d \cdot \mathbf{r} - \mathbf{l}_d \cdot \mathbf{r})} \tag{3.21}$$

does not depend on z. It follows that $k_{d_z} = l_{d_z}$. We will also assume that $k_{d_y} = l_{d_y} = 0$ (the scattering plane corresponds to the *X0Z* plane). Using Maxwell's equations for fields of type (3.2), we can write:

$$\mathbf{H}^*_{l_d} = \frac{c}{\omega}(\mathbf{l}_d \times \mathbf{E}^*_{l_d}), \quad \mathbf{H}_{k_d} = \frac{c}{\omega}(\mathbf{k}_d \times \mathbf{E}_{k_d}), \tag{3.22}$$

then, taking into account **k** = **k***, we obtain the relation

$$\lim_{S\to\infty} \frac{c}{4\pi} \int_S (\mathbf{E}_{k_d} \times \mathbf{H}^*_{l_d} \times \mathbf{E}^*_{l_d} \times \mathbf{H}_{k_d})\mathbf{n}\, dS =$$

$$= \frac{c}{4\pi}\Big[(\mathbf{U}^*_{l_d} \cdot \mathbf{U}_{k_d})(\mathbf{l}_d \cdot \mathbf{n} + \mathbf{k}_d \cdot \mathbf{n}) - (\mathbf{U}_{k_d} \cdot \mathbf{l}_d)(\mathbf{U}^*_{l_d} \cdot \mathbf{n}) -$$

$$- (\mathbf{U}^*_{l_d} \cdot \mathbf{k}_d)(\mathbf{U}_{k_d} \cdot \mathbf{n}) \Big] \cdot \frac{c}{\omega} \lim_{S\to\infty} \int_S e^{i(\mathbf{k}_d \cdot \mathbf{r} - \mathbf{l}_d \cdot \mathbf{r})} dS = \tag{3.23}$$

$$= N_{k_d, l_d} \cdot \delta(k_{d_x} - l_{d_x}) \cdot Y_0,$$

where

$$\delta(k_{d_x} - l_{d_x}) \frac{1}{2\pi} \lim_{x_0\to\infty} \int_{-x_0}^{x_0} e^{i(k_{d_x} - l_{d_x})x} dx \tag{3.24}$$

is the delta function [159], Y_0 are the dimensions of the area occupied by the field sources along the *Y* axis.

The total field in the section $S(z = z_0)$ is equal to the sum of the fields

$$\mathbf{E} = \mathbf{E}_i + \mathbf{E}' + \mathbf{E}'' + \mathbf{E}''', \tag{3.25}$$

where $\mathbf{E}_i$ is the field of the incident light wave,

$\mathbf{E}'$ is the field scattered forward by sources in the volume V_1,

$\mathbf{E}''$ is the field scattered back by sources in the volume V_2,

$\mathbf{E}''' = \dfrac{4\pi}{i\omega\varepsilon}\big(\mathbf{j}_{eq} \cdot \mathbf{z}^0\big)\mathbf{z}^0$ is the field that compensates for possible

discontinuities in the longitudinal components of the electric field caused by the introduction of the $z = z_0$ cross section [158]. To find the fields $\mathbf{E}'$ and $\mathbf{E}''$, we apply the Lorentz lemma in the form (3.19) to the field

$$\mathbf{E}_1 = \sum_{d=1}^{2} \left(\int_{k_{d_z}<0} C_{k_d} \cdot \mathbf{E}_{k_d} dk_{d_x} + \int_{k_{d_z}>0} C_{k_d} \cdot \mathbf{E}_{k_d} dk_{d_x} \right), \quad \mathbf{H}_1, \qquad (3.26)$$

emitted by sources in the volumes V_1 and V_2 and being a superposition of plane waves. The first integral in (3.26) corresponds to waves propagating from right to left, and the second from left to right through the surface S_0. As complex-conjugate fields $\mathbf{E}_2^*$ and $\mathbf{H}_2^*$ appearing in the lemma, we choose a field $\mathbf{E}_{k_d}^* = \mathbf{U}_{k_d}^* \cdot e^{-i(k_d \cdot \mathbf{r} - \alpha t)}$, $\mathbf{H}_{k_d}^*$, satisfying the Maxwell equation without external currents. Taking into account that the plane waves diverge from the region occupied by the field sources, we obtain expressions for the coefficients of the expansion of C_{k_d} entering in (3.26). Taking into account the expression (3.23), we have:

$$\begin{cases} C_{k_d} = -\dfrac{1}{Y_0 \cdot N_{k_d}} \int_{V_1} \mathbf{j}_{eq} \cdot \mathbf{E}_{k_d}^* dV, \quad k_{d_z} > 0, \\[4mm] C_{k_d} = -\dfrac{1}{Y_0 \cdot N_{k_d}} \int_{V_2} \mathbf{j}_{eq} \cdot \mathbf{E}_{k_d}^* dV, \quad k_{d_z} < 0, \end{cases} \qquad (3.27)$$

where

$$N_{k_d} = N_{k_d,k_d} = \frac{c}{2} n_d \cdot \left[2|U_{k_d}|^2 (\mathbf{m}_d \cdot \mathbf{n}) - (\mathbf{m}_d \cdot \mathbf{U}_{k_d})(\mathbf{U}_{k_d}^* \cdot \mathbf{n}) - \right.$$
$$\left. - (\mathbf{m}_d \cdot \mathbf{U}_{k_d}^*)(\mathbf{U}_{k_d} \cdot \mathbf{n}) \right], \qquad (3.28)$$

$\mathbf{m}_d$ is the unit vector of the wave normal: $\mathbf{k}_d = k_d \cdot \mathbf{m}_d = \dfrac{2\pi}{\lambda_0} n_d \cdot \mathbf{m}_d$.

When light is diffracted by sound, the sources of the field are themselves caused by the incident light wave (see 3.9). The main contribution to the scattering field is made by the forward-scattered wave, and the contribution of the field scattered back, due to the corresponding phasing of equivalent currents, is negligible and can be neglected. In what follows we shall also neglect the field $\mathbf{E}'''$ in comparison with $\mathbf{E}'$. Thus, the required field in a certain section $S(z = z_0)$ is described by an integral equation:

$$E = E_i + \sum_{d=1}^{2} \int_{k_{d_z}>0} \left(-\frac{E_{k_d}}{N_{k_d}} \right) \left(\lim_{x_0 \to \infty} \int_{-\frac{L}{2}}^{z_0} \int_{-x_0}^{x_0} j_{eq} \cdot E_{k_d}^* \, dx dz \right) dk_{d_x}, \quad (3.29)$$

where the integration is over all projections k_{d_x} of the wave vectors k_d for which $k_{d_z} > 0$. Substituting (3.9) into (3.29) and using (3.2), we obtain the final form of the integral equation for the diffracted field:

$$E = E_i - \sum_{d=1}^{2} \frac{i\omega}{4\pi} \int_{k_{d_z}>0} \frac{U_{k_d} \cdot e^{i(k_d \cdot r - \omega t)}}{N_{k_d}} \times$$

$$\times \left[\lim_{x_0 \to \infty} \int_{-\frac{L}{2}}^{z_0} \int_{-x_0}^{x_0} (\widehat{\varepsilon^0} \cdot \widehat{p} \cdot \widehat{\xi} \cdot \widehat{\varepsilon^0} E) U_{k_d}^* e^{-i(k_d \cdot r - \omega t)} \, dx dz \right] dk_{d_x}. \quad (3.30)$$

The concrete form of the integral equation (3.30) depends on the choice of the geometry of the acousto-optical interaction, i.e. on the transformation matrix α_{ij}, on the types of interacting light and sound waves.

3.1.2. Solution of the integral equation

We seek the solution of the integral equation (3.30) by the method of successive approximations:

$$E = \sum_{n=0}^{\infty} E^{(n)}. \quad (3.31)$$

We assume that in the scattering plane the incident wave has the form:

$$E^{(0)} = E_i = U_{d_0} \cdot e^{i(k_{d_0} \cdot r - \omega t)} = U_{d_0} \cdot e^{\left[i\frac{2\pi}{\lambda_0} n_d^0 (\cos\theta_0 \cdot z - \sin\theta_0 \cdot x) - \omega t \right]}, \quad d_0 = 1, 2. \quad (3.32)$$

This wave is taken as the zero approximation. For the n-th approximation, after the corresponding transformations [160], one can find:

$$\mathbf{E}^{(n)} = \sum_{d_0} \sum_{d_1} \cdots \sum_{d_n} \left(-\frac{i}{2}\right)^n \cdot \int_{k_{d_n}} \mathbf{U}_{k_{d_n}} \cdot e^{i(\mathbf{k}_{d_n} \cdot \mathbf{r} - \omega_n t)} \times$$

$$\times \int_{k_{d_{n-1}}} \cdots \int_{k_{d_1}} \prod_{m=0}^{n-1} \chi_{d_{m+1}d_m}^{(m)} \cdot e^{-i(\Delta k_{d_x}^{m+1} \mp q_x)^{y \cdot t}} \cdot F(\Delta k_{d_x}^{m+1} \mp q_x) \times \qquad (3.33)$$

$$\times \int_{-1}^{z'_{m+1}} e^{-i\frac{L}{2}(\Delta k_{d_z}^{m+1} \mp q_z)z'_m} \, dz'_m \, dk_{d_{m+1}x} \, .$$

Here the following notation is introduced:

$$\chi_{d_{m+1}d_m}^{(m)} = \frac{\mathbf{U}_{k_{d_{m+1}}}^* \cdot \varepsilon_{k_{d_{m+1}}} \cdot i \cdot P_{ijfp} \cdot \xi_{fp}^0 \cdot \varepsilon_{jk_{d_m}} \cdot \mathbf{U}_{k_{d_m}}}{N_{k_{d_{m+1}}}} \cdot \frac{\omega L}{4\pi}, \qquad (3.34)$$

$\mathbf{U}_{k_{d_{m+1}}}^*$, $\mathbf{U}_{k_{d_m}}$ are the complex amplitudes of the electric field for waves with wave vectors $\mathbf{k}_{d_{m+1}}$ and $\mathbf{k}_{d_m}$; ε_{mn} are the components of the permittivity tensor;

$$F(\Delta k_{d_x}^{m+1} \mp q_x) = \lim_{x_{m+1} \to \infty} \int_{-x_{m+1}}^{x_{m+1}} A(\mathbf{r}'_m) e^{-i(\Delta k_{d_x}^{m+1} \mp q_x) \cdot x_m} \, dx_m \qquad (3.35)$$

is the instant Fourier spectrum of the ultrasonic disturbance $A(\mathbf{r}'_m)$;

$$\Delta k_{d_x}^{m+1} = \frac{2\pi}{\lambda_0} (n_d^{m+1} \cdot \sin\theta_{m+1} - n_d^m \cdot \sin\theta_m), \qquad (3.36)$$

$$\Delta k_{d_z}^{m+1} = \frac{2\pi}{\lambda_0} (n_d^{m+1} \cdot \cos\theta_{m+1} - n_d^m \cdot \cos\theta_m), \qquad (3.37)$$

q_x and q_z are the projections of the wave vector of the elastic wave on the X and Z axes; z'_m is some current coordinate that satisfies the condition $-1 \le z'_m \le 1$.

The choice of the sign $(+)$ or $(-)$ in the expressions (3.33) and (3.35) at a certain stage of the scattering process is related to the fulfillment of the momentum conservation law for incident and diffracted light waves. This choice is due to the fact that real displacements in the local regions of the crystal are described by the real part of expression (3.3).

Based on the essence of this method, the diffraction of light can be represented as the result of successive scattering events into different diffraction orders occurring in accordance with the types of light and sound waves. The summation in (3.33) is over all possible types of light modes with indices d_n arising at each scattering stage. For each m $\chi^{(m)}_{d_{m+1}d_m}$ is the matrix of the relative scattering coefficients

$$\begin{pmatrix} \chi^{(m)}_{11} & \chi^{(m)}_{12} \\ \chi^{(m)}_{21} & \chi^{(m)}_{22} \end{pmatrix}, \tag{3.38}$$

elements of which characterize their type of acousto-optical diffraction at the frequency ω_n, determined by the relation

$$\omega_n = \omega + k\Omega, \tag{3.39}$$

where

$$k = \begin{cases} \pm 1, \ \pm 3, \ ..., \ \pm(2m+1) & \text{at } n=2m+1, \\ 0, \ \pm 2, \ ..., \ \pm 2m & \text{at } n=2m, \ m=0, 1, 2, \ .... \end{cases}$$

It follows from expression (3.33) that the largest contribution to the diffracted field is made by the spectral components of the ultrasonic perturbation for which the synchronism conditions are satisfied:

$$\begin{cases} \Delta k^{m+1}_{d_x} \mp q_x = 0, \\ \Delta k^{m+1}_{d_z} \mp q_z = 0. \end{cases} \tag{3.40}$$

With allowance for (3.40) and solution (3.31), (3.33), a certain scattering diagram can be put in correspondence, where $\prod_{m=0}^{n-1} \chi^{(m)}_{d_{m+1}d_m}$ determines the branch of this diagram, which contributes to the diffraction order. The estimate of the terms of the series (3.33) for each of the branches of the scattering diagram shows that $|E^{(n)}_m| \leqslant M_m \cdot \dfrac{(\chi m)^n}{n!}$, where M_m, χ_m are some constant values. This ensures the convergence of the series (3.31). Thus, the field in the k-th diffraction order can be represented as the sum of the fields taking into account the contributions from each subsequent interaction order, and both the amplitude and the polarization of these additives will be determined by the coefficients $\chi^{(m)}_{d_{m+1}d_m}$, depending on the choice of the scattering geometry and the type of acoustic wave.

3.1.3. Anisotropic diffraction of light by a slow shear wave in a TeO$_2$ crystal at a constant amplitude of an ultrasonic perturbation

We choose the coordinate system X, Y, Z of the problem in question in such a way that it is related to the crystal-physical coordinate system of the crystal by the transformation matrix

$$\alpha_{ij} = \begin{pmatrix} \cos\varphi & \sin\varphi & 0 \\ -\cos\alpha \cdot \sin\varphi & \cos\alpha \cdot \cos\varphi & -\sin\alpha \\ -\sin\alpha \cdot \sin\varphi & \sin\alpha \cdot \cos\varphi & \cos\alpha \end{pmatrix}, \qquad (3.41)$$

where $\varphi = 45°$, α is a small angle characterizing the deviation of the wave vector of the light wave from $(\bar{1}10)$ the plane of the crystal. With this orientation of the scattering plane (Fig. 3.2), the X axis corresponds to the [110] direction of the crystal, and the Z axis makes an angle α with the [001] direction. We assume that along the [110] direction of the crystal an ultrasonic wave propagates with a wave vector **q** whose polarization vector is directed along the $[\bar{1}10]$ axis, this case corresponds to the propagation of a slow shear wave in a TeO$_2$ crystal.

The choice of this scattering geometry is related to the real task of forming an image of an ultrasonic wave modulated in amplitude, which is explained by the optical circuit (Fig. 3.2).

Restricting ourselves to the case of a constant amplitude of the ultrasonic perturbation $A(x) = A_0$, from (3.35) we find:

$$F(\Delta k_{d_x}^{m+1} \mp q_x) = F(k_{d_x}^{m+1} - k_{d_x}^m \mp q_x) =$$

$$= 2\pi A_0 \cdot \delta\left[\frac{2\pi}{\lambda_0}\left(\pm\frac{\lambda_0}{\Lambda} - n_d^{m+1}\cdot\sin\theta_{m+1} + n_d^m\cdot\sin\theta_m\right)\right], \qquad (3.42)$$

where $\delta(x)$ is the Dirac delta function (3.16), $d = 1, 2$. For positive values of the angles θ_m we take the angles measured from the Z axis clockwise in the $X0Z$ plane (Fig. 3.2). In the actual device, there is always a small component $q_z = q \cdot \sin\beta$ associated with the angular divergence of the elastic wave, which will determine the bandwidth of the acousto-optical interaction and whose presence will be taken into account in further calculations.

Solving the system of material equations (3.1), taking into account the smallness of the angles θ_m and α, one can obtain expressions for

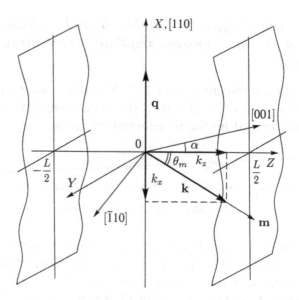

Fig. 3.2. Geometry of broadband diffraction of light on a slow shear wave in a TeO$_2$ crystal.

the refractive indices of two elliptically polarized waves propagating along the *m* direction near the optical axis of the gyrotropic crystal:

$$n_{1,2}^{(m)} = n_o \left[1 \mp \sqrt{a^2 + \tilde{c}^2 \cdot (\theta_m^2 + \alpha^2)^2} + \tilde{c} \cdot (\theta_m^2 + \alpha^2) \right]. \tag{3.43}$$

Here the following notations are introduced:

$$a = \frac{G_{33} \cdot n_o^2}{2}, \quad \tilde{c} = \frac{(n_e^2 - n_o^2)}{4n_e^2}, \tag{3.44}$$

where n_o, n_e are the refractive indices for the ordinary and extraordinary light waves, G_{33} is the component of the pseudotensor of the gyration in the [001] direction of the TeO$_2$ crystal. Since $|G| \simeq 10^{-4} \div 10^{-5}$, then at α, $\theta \simeq 10^{-2}$ we do not take into account the changes in G_{33} for small rotations of the propagation direction of the light wave with respect to the optical axis of the crystal.

From equations (3.1), taking into account the transformation matrix (3.41), we find expressions for the projections of the electric field vectors on the axis of the chosen coordinate system *X, Y, Z*. For complex amplitudes of the electric field of light waves corresponding to the refractive indices $n_d^{(m)}$,

$$\mathbf{U}_{k_d}(\theta_m, \alpha) = U_{x_d}^{(m)} \cdot \mathbf{x}^0 + U_{y_d}^{(m)} \cdot \mathbf{y}^0 +$$

$$+ \left[\frac{n_o^2}{n_e^2} \cdot \theta_m \cdot U_{x_d}^{(m)} + \left(1 - \frac{n_o^2}{n_e^2} \right) \alpha \cdot U_{y_d}^{(m)} \right] \mathbf{z}^0, \tag{3.45}$$

where $\mathbf{x}^0$, $\mathbf{y}^0$, $\mathbf{z}^0$ are the unit vectors of the X, Y, Z coordinate system; $U_{jd} = \eta_{jk} \cdot D_k^{(d)}$, η_{ij} is the tensor of the dielectric impermeability of the medium in the coordinate system associated with the propagating light wave.

The case $d = 1$ corresponds to a fast light wave with a refractive index n_1, while the projections of the electric field vectors can be represented as a matrix

$$U_{j_1} = \frac{D^{(1)}}{n_o^2} \cdot \begin{pmatrix} \cos\varphi_m + i \cdot (\rho_m + \Delta\rho) \cdot \sin\varphi_m \\ \sin\varphi_m + i \cdot (\rho_m + \Delta\rho) \cdot \cos\varphi_m \\ -i \left[\left(1 - \frac{n_o^2}{n_e^2} \right) \cdot \rho_m + \Delta\rho \right] (\alpha \cdot \cos\varphi_m + \theta \cdot \sin\varphi_m) \end{pmatrix}. \tag{3.46}$$

For $d = 2$, which corresponds to a slow light wave with a refractive index n_2,

$$U_{j_2} = \frac{D^{(2)}}{n_o^2} \cdot \begin{pmatrix} -[\sin\varphi_m + i \cdot (\rho_m - \Delta\rho) \cdot \cos\varphi_m] \\ \cos\varphi_m - i \cdot (\rho_m - \Delta\rho) \cdot \sin\varphi_m \\ \left(1 - \frac{n_o^2}{n_e^2} \right) \cdot [\theta \cdot \sin\varphi_m - \alpha \cdot \cos\varphi_m] + \\ + i \cdot \rho_m \cdot (\alpha \cdot \sin\varphi_m + \theta \cdot \cos\varphi_m)] \end{pmatrix}, \tag{3.47}$$

where $D^{(1)}$ and $D^{(2)}$ are the amplitudes of the electric induction vectors for fast and slow light waves in the absence of gyrotropy,

$$\Delta\rho = G_{33} \cdot n_o^2, \tag{3.48}$$

$$\rho_m = [1 + b^2 \cdot (\theta_m^2 + \alpha^2)^2]^{\frac{1}{2}} - b \cdot (\theta_m^2 + \alpha^2) \tag{3.49}$$

is the ellipticity of light waves,

$$b = \frac{(n_e^2 - n_o^2)}{2G_{33} \cdot n_o^2 \cdot n_e^2}. \tag{3.50}$$

The angle φ_m is the angle between the displacement vector of a fast light wave and the X axis. It characterizes the slope of the major axis of the ellipse along which the ends of the induction vectors of an elliptically polarized fast light wave propagate along the direction $\mathbf{m}$ in accordance with the geometry of Fig. 3.2, and is determined by the relation

$$a \cdot \sin \varphi_m = \theta_m \cdot \cos \varphi_m. \tag{3.51}$$

Figure 3.3 is a plot of $\rho(\theta, \alpha)$ for light waves propagating along the $\mathbf{m}$ direction near the [001] optical axis of the TeO$_2$ crystal. The dependence was constructed under the following conditions: $\lambda_0 = 510.6$ nm; $n_o = 2.314$; $n_e = 2.476$; $G_{33} = 3.745 \cdot 10^{-5}$. The angles θ and α are measured in degrees. Figure 3.3 shows that at angles θ, $\alpha \geq 6°$, the ellipticity of the light waves can practically be neglected. Figure 3.4 shows a schematic view of polarization ellipses for a fast light wave propagating at angles θ_m and α to the optical axis of the crystal in question. The direction of rotation of the displacement vector is counterclockwise (left). For a slow light wave, the main axes of the polarization ellipses will be rotated 90° relative to the main axes for a fast light wave, and the direction of rotation of the displacement vector changes to the opposite (right) axis.

When light propagates along the optical axis of the crystal, $\alpha = 0$, $\theta_m = 0$, the light waves will be circularly polarized with $\rho_m = 1$, $\varphi_m = \pi/4$, and have the direction of rotation of the displacement vector opposite to each other.

Performing a multiplication of the tensors in expression (3.34) with allowance for the transformation matrix (3.41) and the type of ultrasonic wave, and also taking into account the expressions for the complex amplitudes of the electric field (3.46)–(3.47), we find the scattering coefficients characterizing the anisotropic diffraction of light with a change in the exponent refraction and polarization. Taking into account the chosen scattering geometry, the following expressions are obtained in the approximation $|G_{33}| \ll \theta_m$, $\alpha \ll 1$.

$$\chi_{12}^{(m)}(\theta_{m+1}, \theta_m) = \frac{n_o^3 \cdot L}{2 \cdot \lambda_0} \cdot \frac{(p_{11} - p_{12}) \cdot \xi_{12}^0}{(1 + \rho_{m+1}^2) \cdot 2} \cdot (U_{1_1}^* U_{2_2} + U_{2_1}^* U_{1_2}) = \tag{3.52}$$

$$= \chi_0 \cdot K_{12}(\theta_{m+1}, \theta_m),$$

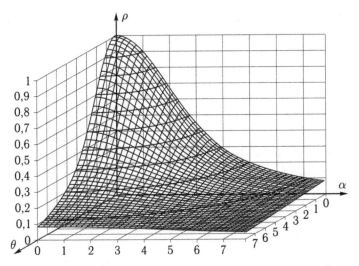

Fig. 3.3. Dependence of the ellipticity of the light wave $\rho(\theta, \alpha)$ on the direction of light propagation near the optical axis of the TeO_2 crystal.

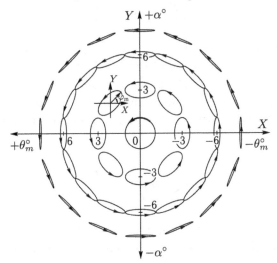

Fig. 3.4. Schematic view of polarization ellipses for a fast diffracted light wave near the optical axis of the TeO_2 crystal.

where

$$\chi_0 = \frac{\pi \cdot A_0 \cdot \xi^0 n_o^3 \cdot L \cdot (p_{11} - p_{12})}{2 \cdot \lambda_0} \cdot \frac{D^{(2)}}{D^{(1)}} \qquad (3.53)$$

is the modulation index, ξ^0 is the normalized deformation of the medium; In what follows, we assume, unless otherwise stated, that $D^{(1)} = D^{(2)}$, (*) is the complex conjugate,

$$K_{12}(\theta_{m+1},\ \theta_m) = \frac{1}{(1+\rho_{m+1}^2)} \cdot [(1+\rho_{m+1}\cdot\rho_m)\cdot\cos(\varphi_m+\varphi_{m+1}) - \\ -i\cdot(\rho_{m+1}+\rho_m)\cdot\sin(\varphi_m+\varphi_{m+1})]. \tag{3.54}$$

In this expression, and further on, the angles in parentheses are read from right to left. Thus, for (3.54), the angle θ_m corresponds to the light wave before scattering, and θ_{m+1} – after scattering.

For backscattering we have:

$$\chi_{21}^{(m)}(\theta_{m+1},\ \theta_m) = \frac{n_o^3 \cdot L}{2\lambda_0} \cdot \frac{(p_{11}-p_{12})\cdot\xi_{12}^0}{2\cdot(1+\rho_{m+1}^2)} \cdot (U_{1_2}^* \cdot U_{2_1} + U_{2_2}^* \cdot U_{1_1}) = \\ = \chi_0 \cdot K_{21}(\theta_{m+1},\ \theta_m), \tag{3.55}$$

$$K_{21}(\theta_{m+1},\ \theta_m) = K_{12}^*(\theta_{m+1},\ \theta_m). \tag{3.56}$$

We assume that the sound wave is excited by a piezotransducer of width L with a uniform amplitude and phase distribution of the field exciting the sound field along the aperture, then the laws of conservation of energy and momentum during the diffraction of light by sound are possible within the angular spectrum of the emitter of elastic waves,

$$\Upsilon(\beta) = \sin\left(\frac{\pi\cdot f\cdot L}{2\cdot v}\cdot\beta\right) \Big/ \left(\frac{\pi\cdot f\cdot L}{2\cdot v}\cdot\beta\right), \tag{3.57}$$

where β is the angle of deviation of the wave vector of the elastic wave from the direction of X. In this case we can write:

$$q_z - \Delta k_{dz}^{(m+1)} = \frac{2\pi}{\lambda_0}\left(\frac{\lambda_0\cdot f}{v}\sin\beta_{m+1,m} - n_d^{(m+1)}\cdot\cos\theta_{m+1} + \\ +n_d^{(m)}\cdot\cos\theta_m\right) = \frac{2\pi}{\lambda_0}\eta_{d^{m+1},d^m}^{(m+1)}(\alpha,\ \theta_{m+1},\ \theta_m). \tag{3.58}$$

Expression (3.42) is transformed to the form

$$F(\Delta k_{d_x}^{m+1} \mp q_x) = 2\pi A_0 \cdot \Upsilon(\beta_{m+1,m}) \times$$

$$\times \delta \left[\frac{2\pi}{\lambda_0} \left(\pm \frac{\lambda_0 \cdot f}{v} \cdot \cos \beta_{m+1,m} - n_d^{(m+1)} \cdot \sin \theta_{m+1} + n_d^{(m)} \cdot \sin \theta_m \right) \right]. \qquad (3.59)$$

The angles θ_m and θ_{m+1} are determined from the conditions for the conservation of momentum for incident and diffracted light waves during anisotropic diffraction of light by sound:

$$\begin{cases} n_d^{(m+1)} \cdot \cos \theta_{m+1} - n_d^{(m)} \cdot \cos \theta_m = \dfrac{\lambda_0 \cdot f}{v} \cdot \sin \beta_{m+1,m}, \\[2mm] n_d^{(m)} \cdot \sin \theta_m - n_d^{(m+1)} \cdot \sin \theta_{m+1} = \dfrac{\lambda_0 \cdot f}{v} \cdot \cos \beta_{m+1,m}, \\[2mm] n_d^{(m,m+1)} = n_o \left[1 \mp \sqrt{a^2 + \tilde{c}^2 \cdot (\theta_{m,m+1}^2 + \alpha^2)^2} + \tilde{c} \cdot (\theta_{m,m+1}^2 + \alpha^2) \right], \end{cases} \qquad (3.60)$$

where $d = 1, 2$.

In the case of anisotropic diffraction in a TeO_2 crystal, in the sequel we shall only take into account scattering events with $\eta_{12}(\alpha, \theta_{m+1}, \theta_m)$ and $\eta_{21}(\alpha, \theta_{m+1}, \theta_m)$ at which transitions from one surface of the refractive index to the other are carried out. Performing the expansion of the functions η_{12}, η_{21}, $n_2^{(0)}$ and $n_1^{(1)}$ in a series in the neighbourhood of the point $\alpha = 0$, we find α, θ for small angles

$$\eta_{12}(\alpha, \theta_{m+1}, \theta_m) \approx \eta_{12}(0, \theta_{m+1}, \theta_m)|_{\alpha=0} +$$
$$+ n_o \cdot \tilde{c} \cdot [\gamma(\theta_{m+1}) + \gamma(\theta_m)]|_{\alpha=0} \cdot \alpha^2, \qquad (3.61)$$
$$\eta_{21}(\alpha, \theta_{m+1}, \theta_m) \approx \eta_{12}(0, \theta_{m+1}, \theta_m)|_{\alpha=0} +$$
$$- n_o \cdot \tilde{c} \cdot [\gamma(\theta_{m+1}) + \gamma(\theta_m)]|_{\alpha=0} \cdot \alpha^2,$$

where

$$\gamma(\theta) = b \cdot \theta^2 \cdot (1 + b^2 \cdot \theta^4)^{-\frac{1}{2}}. \qquad (3.62)$$

We choose a slow light wave with a refractive index $n_2^{(0)}$ as the incident light. This case corresponds to broadband acousto-optic diffraction in a paratellurite crystal. A slow light wave is a wave having a higher refractive index, i.e. we put

$$\mathbf{E}_i = \mathbf{U}_{k_2}(\theta_0) \cdot e^{i(\mathbf{k}_0 \cdot \mathbf{r} - \omega t)}. \qquad (3.63)$$

Then for a field in the +1 order on the output aperture of the AOM with the length of light interaction with the sound $z = L$ (the boundary of the ultrasonic column) and in the first-order approximation of the interaction, we can write:

$$\mathbf{E}_{+1}^{(1)} = -\frac{i}{2}\int\limits_{k_{1z}>0}\mathbf{U}_{k_1}(\theta_1)\cdot\chi_0\cdot\Upsilon(\beta_{1,0})\cdot K_{12}(\theta_1,\ \theta_0)\cdot e^{i[\mathbf{k}_1\cdot\mathbf{r}-(\omega+\Omega)t]} \times$$

$$\times\delta(q_x - \Delta k_{1x}^{(1)})\cdot e^{-i(\Delta k_{1x}^{(1)}-q_x)\cdot vt}\cdot\int\limits_{-1}^{z_0'}e^{-i\frac{L}{2}(q_z-\Delta k_{1z}^{(1)})\cdot z_0'}dz_0'dk_{1x} =$$

$$= -\frac{i}{2}\mathbf{U}_{k_1}(\theta_1)\cdot\chi_0\cdot\Upsilon(\beta_{1,0})\cdot K_{12}(\theta_1,\ \theta_0)\cdot e^{i[\mathbf{k}_1'\cdot\mathbf{r}-(\omega+\Omega)t]} \times$$

$$\times\int\limits_{-1}^{z_0'}e^{i\frac{\pi L}{\lambda_0}\eta_{12}(\alpha,\ \theta_1,\ \theta_0)\cdot z_0'}dz_0',$$

(3.64)

$$\mathbf{k}_1'\cdot\mathbf{r} = \mathbf{k}_0\cdot\mathbf{r} + \Delta k_{1z}\cdot z + \frac{2\pi\cdot f}{v}\cdot\cos\beta\cdot x \approx \mathbf{k}_0\cdot\mathbf{r} + \Delta k_{1z}\cdot z + \frac{2\pi\cdot f}{v}\cdot x,$$

(3.65)

$$\mathbf{k}_0\cdot\mathbf{r} = \frac{2\pi}{\lambda_0}\cdot n_2^{(0)}(\alpha,\ \theta_0)\cdot(\cos\theta_0\cdot z - \sin\theta_0\cdot x) \approx$$

$$\approx (\mathbf{k}_0\cdot\mathbf{r})_{\alpha=0} + \frac{2\pi}{\lambda_0}\cdot n_o\cdot\tilde{c}\cdot[1+\gamma(\theta_0)]\alpha^2\cdot z.$$

(3.66)

We set the initial conditions of the diffraction problem in such a way that for the selected ultrasound carrier frequency $f = f'$, $\theta_0 = \theta_0'$, $\theta_1 = \theta_1'$, where θ_0' and θ_1' are determined from equations (3.60) with $\alpha = 0$, $\beta_{1,0} = 0$.

In this case, the following relation must be satisfied: $\eta_{12}(0,\theta_1',\theta_0') = 0$.

Then, in the first-order approximation, the field at the $+1$ diffraction order at the boundary of the ultrasonic column:

$$\mathbf{E}_{+1}^{(1)} = -i\mathbf{U}_{k_1}(\alpha,\ \theta_1')\cdot\chi_0\cdot K_{12}(\theta_1',\ \theta_0')\cdot e^{i[\mathbf{k}_1'\cdot\mathbf{r}-(\omega+\Omega)t]} \times$$

$$\times\sin c\left\{\frac{L}{\lambda_0}n_o\cdot\tilde{c}\cdot[\gamma(\theta_1')+\gamma(\theta_0')]\cdot\alpha^2\right\},$$

(3.67)

where $\sin c(x) = \dfrac{\sin\pi x}{\pi x}$,

$$\mathbf{k}_1'\cdot\mathbf{r} = (\mathbf{k}_0\cdot\mathbf{r})_{\alpha=0} + \frac{2\pi L}{\lambda_0}n_o\cdot\tilde{c}\cdot[1-\gamma(\theta_1')]\cdot\alpha^2 + \frac{2\pi\cdot f}{v}\cdot x,$$

(3.68)

$$(\mathbf{k}_0 \cdot \mathbf{r})_{\alpha=0} = \frac{2\pi}{\lambda_0} \cdot n_2(\theta_0') \cdot (\cos\theta_0' \cdot z_0 - \sin\theta_0' \cdot x_0) \tag{3.69}$$

is the initial phase of the light wave at the input aperture of the AOM.

Taking into account the second-order interaction, we can determine the field in the second diffraction order $\mathbf{E}_{+2}^{(2)}$, the correction to the zero order $\mathbf{E}_0^{(2)}$, etc. In particular, the field in the +1st order of diffraction in the third-order approximation of the interaction has the form:

$$\mathbf{E}_{+1}^{(3)} = -i\mathbf{U}_{k_1} \cdot \chi_0 \cdot K_{12}(\theta_1', \theta_0') \cdot e^{i[\mathbf{k}_1' \cdot \mathbf{r} - (\omega+\Omega)t]} \times$$

$$\times \left\{ \mathrm{sinc}\,\frac{L}{\lambda_0} \eta_{12}^{(1)}(\theta_1', \theta_0') - \frac{\chi_0^2}{2^3}\left[\Upsilon^2(\beta_{2,1}) \cdot K_{12}(\theta_1', \theta_2) \cdot K_{21}(\theta_2, \theta_1') \times \right.\right.$$

$$\times \int_{-1}^{1}\int_{-1}^{z_1'}\int_{-1}^{z_2'} e^{i\frac{\pi L}{\lambda_0}[\eta_{12}^{(3)}(\theta_1', \theta_2) \cdot z_2' + \eta_{21}^{(2)}(\theta_2,\theta_1') \cdot z_1' + \eta_{12}^{(1)}(\theta_1', \theta_0') \cdot z_0']}\, dz_0' dz_1' dz_2' + \tag{3.70}$$

$$+ K_{12}(\theta_1',\theta_0') \cdot K_{21}(\theta_0',\theta_1') \times$$

$$\left.\left.\times \int_{-1}^{1}\int_{-1}^{z_1'}\int_{-1}^{z_1'} e^{i\frac{\pi L}{\lambda_0}[\eta_{12}^{(3)}(\theta_1', \theta_0') \cdot z_2' + \eta_{21}^{(2)}(\theta_0',\theta_1') \cdot z_1' + \eta_{12}^{(1)}(\theta_1', \theta_0') \cdot z']}\, dz_0' dz_1' dz_2' \right]\right\}.$$

The derivation of expression (3.70) did not take into account the branch of the scattering diagram, which is associated with diffraction in the i-th order. The contribution of this scattering to a field for the +1th order at an angle θ_0' determined by the condition (3.60) will manifest itself only at small interaction lengths at low ultrasound frequencies. For practically important applications, when calculating the field, the contribution from the -1st diffraction order can be neglected. For the specific application of expression (3.70), it should be taken into account that the diffraction of light into the second order, which is taken into account in (3.70), will most effectively occur under the condition: $\eta_{12}^{(1)}\left(\theta_1',\theta_0'\right) = \eta_{21}^{(2)}\left(\theta_2,\theta_1'\right) = \eta_{12}^{(3)}\left(\theta_1',\theta_2\right) = 0$. For our scattering geometry this corresponds to the case $\theta_1' = 0$ and

$$f' = f_{\mathrm{B}} = \frac{v}{\eta_0} n_0 \sqrt{(n_0^2 + n_e^2) \cdot G_{33}}$$ is the frequency of the two-phonon interaction. When creating images, this scattering is harmful.

One of the basic requirements for the AOM for the information display system is its efficiency when the light is deflected into the +1st diffraction order. For this purpose, the carrier frequency and width of the piezo transducer are practically chosen. The wave parameter or the Klein–Cook parameter $Q = \dfrac{2\pi\eta}{n \cdot v} L f^2$ [161] (the ratio of the diffraction angle to the +1st order to the angle of sound divergence) should be much greater than unity (in our case it is approximately equal to 30). To estimate the efficiency and bandwidth of operating frequencies, a linearly frequency-modulated signal approximately corresponding to the video signal band can be fed to the AOM input and the intensity of the light diffracted into the −1st and +2nd diffraction orders is measured. If their value does not exceed 1–2%, then the contribution of these orders in calculating the intensity can be neglected. We will further assume that the frequency $f' \neq f_B$ chosen by us, and, moreover, f_B lies outside the frequency band of the acousto-optical modulator. In this case, the contribution from the second order of diffraction to a field for the +1st order can be neglected and allow only the interaction between the +1st and 0th orders of diffraction.

3.1.4. Anisotropic diffraction with amplitude modulation of the ultrasonic signal

Suppose that the amplitude of the ultrasonic perturbation in the crystal varies according to the law

$$A(\mathbf{r}) = A_0 \left[1 + m_0 \cdot \cos\left(2\pi f_0 \cdot \frac{\mathbf{r}}{\mathbf{v}} \right) \right] = A_0 \cdot [1 + m_0 \cdot \cos(\Phi_0 \cdot x)], \qquad (3.71)$$

where m_0 is the modulation depth, f_0 is the modulation frequency, $\mathbf{v}$ is the velocity vector, and $\mathbf{r}$ is the propagation direction of the ultrasonic disturbance in the medium. We assume that at small angles β of the deviation of the propagation direction from the X axis, the velocity of the ultrasonic wave varies insignificantly.

Substituting (3.64) into (3.27) with allowance for (3.50), we find

$$F(\Delta k_{1x}^1 \mp q_x) = 2\pi A_0 \cdot \Upsilon(\beta) \cdot \left\{ \delta\left(\Delta k_{1x}^1 \mp \frac{2\pi \cdot f}{v} \cdot \cos\beta \right) + \right.$$
$$+ \frac{m_0}{2} \delta\left[\Delta k_{1x}^1 \mp \frac{2\pi}{v}(f - f_0) \cdot \cos\beta \right] + \frac{m_0}{2} \qquad (3.72)$$
$$\left. \delta\left[\Delta k_{1x}^1 \mp \frac{2\pi}{v}(f + f_0) \cdot \cos\beta \right] \right\}.$$

Assuming that the conditions (3.60) for $\alpha = 0$, $\beta = 0$:
$\theta_0 = \theta_0^{(0)}$, $\theta_1 = \theta_0^{(1)}$ are satisfied for the central spectral component of the amplitude-modulated signal, and also taking into account only the interaction between (+1) and (0) diffraction orders, we calculate the field in the (+1) diffraction order at the boundary of the ultrasonic column in the AOM made from TeO_2. The field is calculated according to the scattering diagram shown in Fig. 3.5.

The horizontal arrows on the diagram characterize the partial plane waves propagating at angles $\theta_p^{(m)}$ for $m = 0, 1, 2, 3$ (zero, first, second and third orders of interaction). The inclined arrows denote the possible directions of scattering of these waves by the action of an ultrasonic perturbation (3.71).

It follows from the diagram that in the third-order approximation of the interaction for a signal with three spectral components, there are 27 possible scattering combinations contributing to the diffracted field near +1 order. The spectrum of this field consists of 7 components, differing by the diffraction angles $\theta_p^{(3)}$ and frequencies by the value $\Delta\Omega_p = p \cdot 2\pi \cdot f_0$, where $p = -3, -2, -1, 0, +1, +2, +3$.

The calculation of the diffracted field is performed in accordance with the general solution (3.33) by successively finding the angles $\theta_p^{(m+1)}$ and $\beta_{p,q}^{(m+1)}$, satisfying the system of equations (3.60), in which the following substitution should be made: $\theta_{m+1} = \theta_p^{(m+1)}$, $\theta_m = \theta_q^{(m)}$, $\beta_{m+1,m} = \beta_{p,q}^{(m+1)}$; where $p, q = -3, -2, -1, 0, +1, +2, +3$.

The final angles $\theta_p^{(m+1)}$ obtained in the previous stage are the initial angles in the following interaction order. Simultaneously with the angles $\theta_p^{(m+1)}$, the angles $\beta_{p,q}^{(m+1)}$ and the weight spectral $\Upsilon\left(\beta_{p,q}^{(m+1)}\right)$ coefficients are found from the equations (3.60). Summarizing all the scattering combinations shown in Fig. 3.5, we find the field in

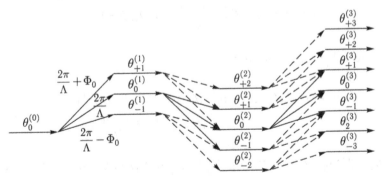

Fig. 3.5. Diagram of scattering by light diffraction in +1st order in the approximation of the third order of interaction.

the +1st diffraction order on the output AOM with allowance for the third-order approximation of the interaction:

$$\mathbf{E}_{+1}^{(3)} = -i \cdot \chi_0 \cdot e^{i\left[(\mathbf{k}_0 \cdot \mathbf{r})_{\alpha=0} - \omega \cdot t + \frac{2\pi \cdot f}{v}(x - v \cdot t) + 2L_0 \cdot \alpha^2\right]} \times$$

$$\times \sum_{p=-3}^{3} C_p(\theta_p^{(3)}) \cdot e^{i\left[p \cdot \frac{2\pi \cdot f_0}{v}(x - v \cdot t) - 2L_0 \cdot \gamma_p^{(3)} \cdot \alpha\right]}, \tag{3.73}$$

$$C_0(\theta_0^{(3)}) = \mathbf{U}_{k_1}(\theta_0^{(3)}) \cdot \left[K_{12}(\theta_0^{(1)}, \theta_0^{(0)}) \cdot \mathrm{sinc}\left[\frac{L_0}{\pi} \cdot \alpha^2 \cdot (\gamma_0^{(1)} + \gamma_0^{(0)})\right] - \right.$$

$$- \frac{\chi_0^2}{8} \left\{ S(0,0,0) + \frac{m_0^2}{4} \cdot [S(0,0,+1) + S(0,0,-1) + S(0,+1,+1) + \right. \tag{3.74}$$

$$\left. \left. + S(0,+1,0) + S(0,-1,0) + S(0,-1,-1)]\right\}\right]$$

$$C_{\pm 1}(\theta_\pm^{(3)}) = \mathbf{U}_{k_1}(\theta_{\pm 1}^{(3)}) \cdot \frac{m_0}{2} \cdot \left[K_{12}(\theta_{\pm 1}^{(1)}, \theta_0^{(0)}) \cdot \Upsilon(\beta_{\pm 1,0}^{(1)}) \times \right.$$

$$\times \mathrm{sinc}\left[\frac{L_0}{\pi} \cdot \alpha^2 \cdot (\gamma_{\pm 1}^{(1)} + \gamma_0^{(0)})\right] - \frac{\chi_0^2}{8} \cdot \{S(\pm 1, \pm 1, \pm 1) + S(\pm 1, \pm 1, 0) + \tag{3.75}$$

$$\left. + S(\pm 1, 0, 0) + \frac{m_0^2}{4} \cdot [S(\pm 1, 0, \pm 1) + S(\pm 1, 0, \mp 1) + S(\pm 1, \pm 2, \pm 1)]\}\right],$$

$$C_{\pm 2}(\theta_{\pm 2}^{(3)}) = -\mathbf{U}_{k_1}(\theta_{\pm 2}^{(3)}) \cdot \frac{\chi_0^2}{8} \cdot \frac{m_0^2}{4} \times$$

$$\times [S(\pm 2, \pm 2, \pm 1) + S(\pm 2, \pm 1, \pm 1) + S(\pm 2, \pm 1, 0)], \tag{3.76}$$

$$C_{\pm 3}(\theta_{\pm 3}^{(3)}) = -\mathbf{U}_{k_1}(\theta_{\pm 3}^{(3)}) \cdot \frac{\chi_0^2}{8} \cdot \frac{m_0^3}{8} \cdot S(\pm 3, \pm 2, \pm 1), \tag{3.77}$$

with the following notations

$$S(p,q,l) = L(\theta_p^{(3)}, \theta_q^{(2)}, \theta_l^{(1)}, \theta_0^{(0)}) \cdot \Upsilon(\beta_{p,q}^{(3)}, \beta_{q,l}^{(2)}, \beta_{l,0}^{(1)}) \cdot$$

$$J(\gamma_p^{(3)}, \gamma_q^{(2)}, \gamma_l^{(1)}, \gamma_0^{(0)}), \tag{3.78}$$

$$L(\theta_p^{(3)}, \theta_q^{(2)}, \theta_l^{(1)}, \theta_0^{(0)}) = K_{12}(\theta_p^{(3)}, \theta_q^{(2)}) \cdot K_{21}(\theta_q^{(2)}, \theta_l^{(1)}) \cdot K_{12}(\theta_l^{(1)}, \theta_0^{(0)}), \tag{3.79}$$

$$\Upsilon(\beta^{(3)}_{p \cdot q}, \beta^{(2)}_{q \cdot l}, \beta^{(1)}_{l \cdot 0}) = \Upsilon(\beta^{(3)}_{p,q}) \cdot \Upsilon(\beta^{(2)}_{q,l}) \cdot \Upsilon(\beta^{(1)}_{l,0}), \tag{3.80}$$

$$J(\gamma^{(3)}_p, \gamma^{(2)}_q, \gamma^{(1)}_l, \gamma^{(0)}_0) =$$

$$= \int_{-1}^{1} \int_{-1}^{z'_2} \int_{-1}^{z'_1} e^{iL_0 \cdot \alpha^2 \cdot [(\gamma^{(3)}_p + \gamma^{(2)}_q) \cdot z'_2 - (\gamma^{(2)}_q + \gamma^{(1)}_l) \cdot z'_1 + (\gamma^{(1)}_l + \gamma^{(0)}_0) + z'_0]} dz'_2 dz'_1 dz'_0, \tag{3.81}$$

$$\gamma^{(m)}_p = \gamma(\theta^{(m)}_p) = b \cdot (\theta^{(m)}_p)^2 \cdot [1 + b^2 \cdot (\theta^{(m)}_p)^4]^{-\frac{1}{2}}, \tag{3.82}$$

$$L_0 = \frac{\pi \cdot L \cdot n_o \cdot \tilde{c}}{\lambda_0} = \frac{\pi \cdot L \cdot n_o \cdot (n_e^2 - n_o^2)}{\lambda_0 \cdot 4 \cdot n_e^2}. \tag{3.83}$$

The amplitudes $U_{k_1}(\theta^{(3)}_p)$ are determined by the expressions (3.46), (3.47). It can be seen from the relations (3.73)–(3.83) that in the general case the amplitudes of the spectral components with positive and negative p indices become not identical. The symmetry of the spectrum with respect to the direction $\theta_1 = \theta^{(1)}_0 = \theta^{(3)}_0$ is violated, and the shape of the spectral envelope depends on the choice of the carrier frequency f and the adjustment of the central spectral component of the spectrum to the Bragg diffraction conditions. The amplitudes of the spectral components with 'parasitic' modulation frequencies ($p = \pm 2, \pm 3$) increase with increasing modulation index χ_0 and strongly depend on the modulation depth m_0. At a small depth of modulation, the contribution of these components to the general field can be neglected. It is not difficult to see, in the case of visualization of images, in order to obtain a good contrast, it is necessary to work with the values $m_0 \simeq 1$, therefore these components can not be neglected at large modulation indices.

In concluding this section, we note that in a real device the field at the input aperture of the acousto-optical modulator is a superposition of plane waves, which is the result of the diffraction of the incoming light beam at the input aperture. In this case, each diffraction field of type (3.70) or (3.73) will correspond to each plane wave. The influence of the acousto-optical modulator on the angular spectrum of the input optical signal can be represented as the action of some, in the general case, nonlinear operator, leading to the transformation of the amplitudes of the spectral components of the signal and the displacement of the wave vectors in accordance with the spectrum of the ultrasonic perturbation. For small modulation indices χ_0, this operator can be regarded as linear and confined to field calculations only by the first order of interaction. Nevertheless, the use of expressions for the diffracted field with allowance for the third order of interaction makes it possible to establish the limits of applicability of the first approximation, and also to solve the problem of nonlinear distortions in the image signal.

3.2. Forming a line image in an acousto-optical system with a pulsed coherent light source

Consider an acousto-optical system for the formation of a line image (Fig. 3.6) consisting of an acousto-optical modulator (AOM), an input cylindrical lens L_1, and an objective lens L_2 and L_3.

   The planes $X'0Z'$ and $Y'0Z'$ of the system are chosen in such a way that in one of them (the plane $X'0Z'$) the light diffracts into the AOM and the image of the line is displayed on the screen E, and in the other, the spatial distribution of the line and its deviation in the plane of the screen. In the focal plane of the lens L_2 there is a diaphragm D, which removes the zero order of diffraction, leaving only the diffracted light beams in the +1st diffraction order.

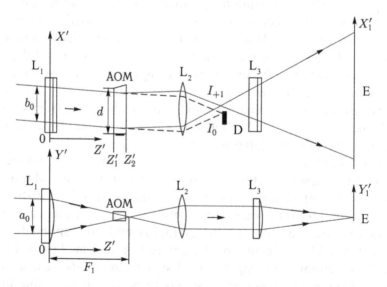

Fig. 3.6. Optical scheme of the imaging system of a line in two mutually perpendicular planes: AOM – acousto-optical modulator; L_1 – input cylindrical lens; L_2, L_3 – the objective lens; E – screen; D – diaphragm; I_0, I_{+1} – the intensity of the incident and diffracted light rays; a_0, b_0 – dimensions of the light beam in two coordinates; F_1 is the focal length of A_1; d is the size of the acousto-optical modulator along the X' axis.

The acousto-optical interaction medium is a TeO_2 crystal. Our consideration will relate to a specific interaction geometry that takes place in a real device (Figs. 3.7 and 3.2), corresponding to the case of broadband anisotropic light diffraction on a slow shear elastic wave. We shall assume that an ultrasonic wave **q** propagates along the [110] direction of the crystal coinciding with the axis X' of the optical system, and the polarization vector of this wave is directed along the axis $[\bar{1}10]$, coinciding with the Y' axis.

The wave vector of the light wave **k** makes an angle θ with its projection onto the (110) plane, and the projection of this vector on the [110] plane makes an angle α with the Z' axis. The angles α and θ are assumed small, then

$$k_{x'}, k_{y'} \ll k; \quad k_{x'} = k_x = -k \cdot \sin\theta \approx -k \cdot \theta; \quad k_{y'} = k \cdot \cos\theta \cdot \sin\alpha \approx k \cdot \alpha, \quad (3.84)$$

where $k = 2\pi/\lambda_0$ is the amplitude of the wave vector of the light wave for free space, $k_{x'}$, $k_{y'}$ are the projections of the wave vector of the light wave on the coordinate axis.

Suppose that a homogeneous plane light wave is incident on the input of the system, the electric field vector of which varies according to the law:

$$\mathbf{E}_i = \mathbf{U}_l \cdot e^{i(\mathbf{k}_0 \cdot \mathbf{r}' - \omega t)} = \tilde{\mathbf{U}}_l \cdot e^{-i\omega t}, \quad (3.85)$$

where

$$\mathbf{U}_l = U_x \cdot \mathbf{x}^0 + U_y \cdot \mathbf{y}^0 + U_z \cdot \mathbf{z}^0 = U_1 \cdot \mathbf{x}^0 + U_2 \cdot \mathbf{y}^0 + U_3 \cdot \mathbf{z}^0, \quad (3.86)$$

$\mathbf{x}^0$, $\mathbf{y}^0$, $\mathbf{z}^0$ are the unit vectors, $\mathbf{k}_0 \cdot \mathbf{r}' = k \cdot (\cos\tilde{\theta}_0 \cdot \cos\tilde{\alpha}_0 \cdot z' - \sin\tilde{\theta}_0 \cdot x' + \cos\tilde{\theta}_0 \times \times \sin\tilde{\alpha}_0 \cdot y')$.

We assume that the angles $\tilde{\theta}$ and $\tilde{\alpha}$ outside the crystal are related to the angles θ and α inside the crystal by the corresponding Snellius law at the media interface. For definiteness, we also assume that the wave (3.85) is an elliptically polarized slow light wave with a refractive index $n_2^{(0)}$ (see (3.43)) for $\alpha = 0$, $\theta = \theta_0$.

The complex amplitude of each component of the electric field $\tilde{U}_l = \tilde{U}_l(x, y, z)$ is a function of the coordinates of points in space that satisfies the scalar wave equation, as well as the Kirchhoff diffraction formulas [162]. In this regard, we will consider the process of image formation from the point of view of the scalar wave theory, which is valid for each Cartesian component of the electric field.

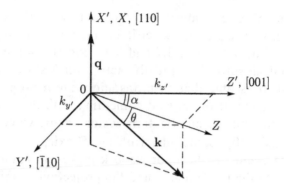

Fig. 3.7. Geometry of acousto-optical interaction.

Suppose that a thin cylindrical lens L_1 having a focal length F_1 is located at the input of the optical system at a distance $z' = z'_1$ from the AOM. We will consider it as a screen with a transmission function

$$g(x'_0, y'_0) = g_1(x'_0) \cdot g_2(y'_0),$$

where

$$g_1(x'_0) = \begin{cases} 1, \text{ at } |x'_0| \leqslant \dfrac{b_0}{2}, \\ 0, \text{ at } |x'_0| > \dfrac{b_0}{2}, \end{cases} \quad g_2(y'_0) = \begin{cases} e^{-i\frac{k}{2F_1}y'_0{}^2}, \text{ at } |y'_0| \leqslant \dfrac{a_0}{2}, \\ 0, \text{ at } |y'_0| > \dfrac{a_0}{2}. \end{cases} \quad (3.87)$$

We represent the complex amplitude of each of the orthogonal components of the light field incident on the lens in the form

$$\tilde{U}_l = U_{l_0} \cdot e^{i(k_{0z'} \cdot z' + k_{0x'} \cdot x' + k_{0y'} \cdot y')} =$$
$$= U_{l_0} \cdot e^{ik(\cos\tilde{\theta}_0 \cdot \cos\tilde{\alpha}_0 \cdot z' - \sin\tilde{\theta}_0 \cdot x' + \cos\tilde{\theta}_0 \cdot \sin\tilde{\alpha}_0 \cdot y')}, \quad (3.88)$$

where $l = 1, 2, 3$.

The field behind the screen at $z' = 0$ is determined by the expression

$$\tilde{U}_l(x'_0, y'_0) = U_{l_0} \cdot e^{ik_{0x'} \cdot x'_0} \cdot g_1(x'_0) \cdot e^{ik_{0y'} \cdot y'_0} \cdot g_2(y'_0). \quad (3.89)$$

Taking into account (3.84), the angular spectrum of this field has the form

$$\Phi_{l_0}(k_{x'},k_{y'})=U_{l_0}\cdot\int_{-\infty}^{\infty}\int_{-\infty}^{\infty}g_1(x_0')\cdot e^{-i(k_{x'}-k_{0x'})\cdot x_0'}\cdot g_2(y_0')\times$$

$$\times e^{-i(k_{y'}-k_{0y'})\cdot y_0'}\,dx_0'\,dy_0'\approx U_{l_0}\cdot\tilde{J}\cdot 2\pi\cdot\delta(k_{x'}-k_{0x'}),\tag{3.90}$$

Taking into account the expressions (3.84) for $\dfrac{k}{2F_1}\cdot\dfrac{a_0^2}{4}\gg 1$, we use the first term of the asymptotic expansion of the integral $\tilde{J}$ in inverse powers of k [152]. For $|\alpha-\alpha_0|\leqslant\dfrac{a_0}{2F_1}$ with:

$$\tilde{J}\approx\sqrt{\frac{2\pi F_1}{k}}\cdot e^{i\left(\frac{k\cdot F_1(\alpha-\alpha_0)^2}{2}+\frac{3\pi}{4}\right)}+o\!\left(k^{-\frac{1}{2}}\right),\tag{3.91}$$

where $\alpha=\dfrac{k_{y'}}{k}$, $\alpha_0=\dfrac{k_{0y'}}{k}$.

The angular spectrum of the light field in the plane $z'=z_1'$, the input plane of the acousto-optical modulator, has the form [163]:

$$\Phi_I(k_{x'},k_{y'},z_1')=\Phi_{l_0}(k_{x'},k_{y'})\cdot e^{iz_1'\cdot\sqrt{k^2-k_{x'}^2-k_{y'}^2}}\approx$$

$$\approx\Phi_{l_0}(k_{x'},k_{y'})\cdot e^{ikz_1'}\cdot e^{-i\frac{(k_{x'}^2+k_{y'}^2)}{2k}\cdot z_1'}\approx U_{l_0}\cdot 2\pi\delta(k_{x'}-k_{0x'})\times\tag{3.92}$$

$$\times e^{i\left(k\cdot z_1'+\frac{3\pi}{4}\right)}\cdot\sqrt{\frac{2\pi F_1}{k}}\cdot e^{-i\frac{(k_{x'}^2+k_{y'}^2)}{2k}\cdot z_1'}\cdot e^{i\frac{(k_{y'}-k_{0y'})^2}{2k}}\cdot F_1.$$

Neglecting the loss of light in the reflection of waves at the air-crystal interface, the field interacting with the ultrasound can be represented as a superposition of plane waves:

$$\tilde{U}_{l_2}(x,y,z')=\frac{1}{(2\pi)^2}\cdot\int_{-\infty}^{\infty}\int_{-\infty}^{\infty}\Phi_{l_2}(k_{2x},k_{2y},z')\cdot e^{i(k_{2x}\cdot x+k_{2y}\cdot y)}\,dk_{2x}\,dk_{2y},\tag{3.93}$$

where $\Phi_{l_2}\left(k_{2x},k_{2y},z'\right)$ is the angular spectrum of plane waves with the refractive index $n_2^{(0)}$,

$$\Phi_{l_2}(k_{2x},k_{2y},z')=2\pi\sqrt{\frac{F_1\cdot\lambda_0}{n_2^{(0)}}}\cdot\delta(k_{2x}-k_{2x}^0)\cdot U_{l_2}\times$$

$$\times e^{i\left(k_2\cdot z'+\frac{3\pi}{4}\right)}\cdot e^{i\left[\frac{(k_{2y}-k_{2y}^0)^2}{2\cdot k_2}\cdot F_1-\frac{(k_{2x}^2+k_{2y}^2)}{2\cdot k_2}\cdot z'\right]},\tag{3.94}$$

$$k_{2y}^0=\frac{2\pi}{\lambda_0}\cdot n_2^{(0)}\cdot\cos\theta_0\cdot\sin\alpha_0,\ k_{2x}^0=-\frac{2\pi}{\lambda_0}\cdot n_2^{(0)}\cdot\sin\theta_0.$$

Here and below, the value F_1 is the equivalent focal length of the lens A_1 with allowance for the refraction of light in the crystal.

As a result of interaction with the ultrasonic field in the AOM, starting from the distance $z' = z_1'$, the angular spectrum of the incident light wave is transformed: $\Phi_{l_2}(k_{2x},k_{2y},z)\to\Phi_{l_1}^{(+1)}(k_{1x},k_{1y},z)$. The light field after the AOM in the +1st diffraction order will be a superposition of diffracted plane waves. Using the general solution of the integral equation (3.31)–(3.39), we can find the total field in the +lth diffraction order on the output aperture of the acousto-optical modulator ($z' = z_2' = z_1' + L$) in the nth order approximation of the interaction. In particular, when the acousto-optical interaction is of low efficiency, we can confine ourselves to the first-order approximation of the interaction. Then for the field in the +1st diffraction order on the output aperture AOM, we find:

$$E_{+1}^{(1)}=\tilde{U}_{l_1}\cdot e^{-i(\omega+\Omega)t},\tag{3.95}$$

where

$$\tilde{U}_{l_1}=\tilde{U}_{x1}\cdot\mathbf{x}^0+\tilde{U}_{y1}\cdot\mathbf{y}^0+\tilde{U}_{z_1}\cdot\mathbf{z}^0.\tag{3.96}$$

Each of the complex amplitudes of the orthogonal components of the light field (3.96) as a result of diffraction is transformed to the form:

$$\tilde{U}_{l_1}(x,y,z)=\frac{1}{(2\pi)^2}\cdot\int_{-\infty}^{\infty}\int_{-\infty}^{\infty}\Phi_{l_1}^{(+1)}(k_{1x},k_{1y},z)\cdot e^{i(k_{1x}\cdot x+k_{1y}\cdot y)}dk_{1x}dk_{1y}=$$

$$=\frac{1}{2\pi}\cdot\sqrt{\frac{F_1\cdot\lambda_0}{n_1^{(1)}}}\cdot e^{i\cdot\frac{3\cdot\pi}{4}}\cdot e^{i(k\cdot z_1+k_1\cdot z)}\cdot e^{-ik\left(\theta_0\cdot x'+\frac{\theta_0^2}{2}\cdot z'\right)}\cdot P(x,y)\times\tag{3.97}$$

$$\times\int_{-\infty}^{+\infty}U_{l_1}^{(+1)}\cdot e^{ik_{1y}\cdot y}\cdot e^{-i\frac{k_{1y}^2}{2k_1}\cdot z'}\cdot e^{i\frac{(k_{1y}-k_{1y}^0)^2}{2\cdot k_1}\cdot F_1}dk_{1y}.$$

For a field in the plane $z' = F_1$, approximately coinciding with the plane of the output aperture AOM $z' = z'_2 = z'_1 + L$, we obtain the expression

$$\tilde{U}_{I_1}(x,y) = \sqrt{\frac{F_1}{\lambda_0}} \cdot e^{i \cdot \frac{3 \cdot \pi}{4}} \cdot e^{i(k \cdot z'_1 + k_1 \cdot L)} \cdot e^{-ik\left[\theta_0 \cdot x' + \frac{(\theta_0^2 - \alpha_0^2)}{2} \cdot F_1\right]} \cdot P(x,y) \times$$

$$\times \int_{-\tilde{a}}^{\tilde{a}} U_{I_1}(\alpha, \theta_p^m) \cdot e^{i \cdot k_1 \cdot \alpha \cdot (y - \alpha_0 \cdot F_1)} d\alpha, \tag{3.98}$$

where $U_{I_1}(\alpha, \theta_p^m)$ are the complex amplitudes of the orthogonal components of the electric field of the diffracted light waves for a fixed value of the angle α. Here and below, the value F_1 is the equivalent focal length of the lens L_1, taking into account the passage of light through the AOM crystal;

$$P(x,y) = P(x) = \begin{cases} 1, & \text{at } |x| \leqslant \dfrac{d}{2}, \\ 0, & \text{at } |x| > \dfrac{d}{2}, \end{cases} \tag{3.99}$$

d is the size of the AOM along the X axis.

We will assume that the amplitude of the ultrasonic signal varies according to the harmonic law (3.71), then substituting the complex amplitudes from (3.73)–(3.75) into (3.98) and restricting ourselves to the first-order approximation, we find an expression for the field in the plane $z' = F_1$ approximately coinciding with the plane of the output aperture AOM $z' = z'_2 = z'_1 + L$.

$$\tilde{U}_{I_1}(x',y',t) = \Theta_0 \cdot P(x') \cdot \chi_0 \cdot e^{i\frac{2\pi}{v} \cdot f \cdot x'} \left\{ \int_{-\tilde{a}}^{\tilde{a}} U_{I_1}(\alpha, \theta_0^{(1)}) \cdot K_{12}(\theta_p^{(1)}, \theta_0^{(0)}, \alpha) \times \right.$$

$$\times \operatorname{sinc}\left[\frac{L_0}{\pi} \cdot (\gamma_0^{(1)} + \gamma_0^{(0)}) \cdot \alpha^2\right] \cdot e^{iv_0(\alpha, y')} d\alpha +$$

$$+ \frac{m_0}{2} \cdot \sum_{p=-1,1} e^{ip \cdot \frac{2\pi \cdot f_0}{v} \cdot (x' - v \cdot t)} \cdot \Upsilon(\beta_{p,0}^{(1)}) \cdot \int_{-\tilde{a}}^{\tilde{a}} \frac{U_{I_1}(\alpha, \theta_p^{(1)}) \cdot}{K_{12}(\theta_p^{(1)}, \theta_0^{(0)})} \times$$

$$\times e^{iv_p(\alpha, y')} \cdot \operatorname{sinc}\left[\frac{L_0}{\pi} \cdot (\gamma_p^{(1)} + \gamma_0^{(0)}) \cdot \alpha^2\right] d\alpha \right\}, \tag{3.100}$$

where

$$v_p(\alpha, y') = \frac{2\pi}{\lambda_0} n_0 \cdot [\alpha \cdot (y' - \alpha_0 \cdot F_1) + L \tilde{c} \cdot (1 - \gamma_p^{(1)}) \cdot \alpha^2], \qquad (3.101)$$

$$\Theta_0 = \sqrt{\frac{F_1}{\lambda_0}} \cdot e^{i\left(\frac{2\pi}{\lambda_0} \cdot R_0 + \frac{\pi}{4}\right)}, \qquad (3.102)$$

$R_0 = (\mathbf{k}_0 \cdot \mathbf{r})|_{\alpha = \alpha_0}$ is the optical path length for the central light beam between the planes $(z' = 0)$ and $(z' = z_1')$, $\tilde{\alpha} = \dfrac{a_0}{2F_1 \cdot n_0}$. In what follows we assume that $P(x') = 1$.

Expression (3.100) is the sum of products of functions that depend on the coordinates x' and y':

$$\tilde{U}_{l_1}(x', y', t) = \Theta_0 \cdot [\overline{f_0(x') \cdot g_0(y')} + $$
$$+ \hat{f}_1(x') \cdot \hat{g}_1(y') \cdot e^{-i\Phi_0 \cdot v \cdot t} + \hat{f}_2(x') \cdot \hat{g}_2(y') \cdot e^{i\Phi_0 \cdot v \cdot t}]. \qquad (3.103)$$

We assume that the image forming optical system is constructed in such a way that in one plane $X'0Z'$ it carries the image of the diffracted field from the output aperture of the AOM to the screen plane with magnification M_1', and in the other $(Y'0Z')$ field transfers the field distribution to $(z' = F_1)$ to the screen with magnification M_2'. In this case, the field in the plane of the screen will look like:

$$\tilde{U}_{l_1}(x_1', y_1', t)|_{z' = z_e} = \Theta_1 \cdot [\hat{F}_0(x_1') \cdot \hat{G}_0(y_1') + $$
$$+ \hat{F}_1(x_1') \cdot \hat{G}_1(y_1') \cdot e^{-i\Phi_0 \cdot v \cdot t} + \hat{F}_2(x_1') \cdot \hat{G}_2(y_1') \cdot e^{i\Phi_0 \cdot v \cdot t}]. \qquad (3.104)$$

We will assume that the size of the lens that builds the image is large enough. Then $\hat{f}(x') = 0$ for $|x'| > d_0$, $\hat{g}(y') = 0$ for $|y'| > d_0$, $2d_0$ is the diameter of the aperture of the objective lens, and also $2\beta_0 d_0 \gg 1$, which, as a rule, is satisfied for a real optical system, where $\tilde{\beta}_0 = \dfrac{k}{2z_s'}$, z_s' is the distance from the objective to its object plane, z_0' is the distance from the lens to the image plane $(z_s'$ and z_0' may not be the same for the planes $X'0Z'$ and $Y'0Z')$. In addition, the functions $\hat{f}(x)$ and $\hat{F}(x)$, $\hat{g}(y)$ and $\hat{G}(y)$ are related by [162]

$$\hat{F}(x_1') = \sqrt{\frac{z_s'}{z_0'}} \cdot e^{i \cdot \frac{\pi \cdot z_s' \cdot (x_1')^2}{\lambda_0 \cdot z_0' \cdot F_2}} \cdot \hat{f}\left(-\frac{z_s' \cdot x_1'}{z_0'}\right) = \frac{e^{i \cdot \frac{\pi \cdot (x_1')^2}{\lambda_0 \cdot M_1' \cdot F_2}}}{\sqrt{M_1'}} \cdot \hat{f}\left(-\frac{x_1'}{M_1'}\right), \qquad (3.105)$$

$$\hat{G}(y_1') = \frac{e^{i \cdot \frac{\pi \cdot z_s' \cdot (x_1')^2}{\lambda_0 \cdot M_2' \cdot F_3}}}{\sqrt{M_2'}} \cdot \hat{g}\left(-\frac{y_1'}{M_2'}\right). \qquad (3.106)$$

Introducing the notation, $\left(-\dfrac{x_1'}{M_1'}\right) = x'$, $\left(-\dfrac{y_2'}{M_2'}\right) = y'$, the field in the plane of the screen (image plane) can be represented as:

$$\tilde{U}_{l_1}(x_1', y_1', t)\big|_{z'=z_e} = \Theta_1 \cdot [\hat{f}_0(x') \cdot \hat{g}_0(y') + \hat{f}_1(x') \cdot \hat{g}_1(y') \cdot e^{-i\Phi_0 \cdot v \cdot t} + $$
$$+ \hat{f}_2(x') \cdot \hat{g}_2(y') \cdot e^{i\Phi_0 \cdot v \cdot t}], \qquad (3.107)$$

where

$$\Theta_1 = \sqrt{\frac{F_1}{\lambda_0 \cdot M_1' \cdot M_2'}} \cdot e^{i\left[\frac{2\pi}{\lambda_0} \cdot R_1(z') + \frac{\pi}{4}\right]} \cdot e^{i\frac{\pi}{\lambda_0}\left(\frac{x_1'^2}{M_1' \cdot F_2} + \frac{y_1'^2}{M_2' \cdot F_3}\right)}. \qquad (3.108)$$

$R_1(z')$ is the total length of the optical path for the central ray from the plane $(z' = 0)$ to $(z' = z_e')$, F_2 and F_3 are the equivalent focal lengths of the lens that builds the line image on the screen for the planes $X'0Z'$ and $Y'0Z'$, respectively.

For instantaneous light intensity in the image plane, we find:

$$I_1(x_1', y_1', t) = \sum_{l=1}^{3} I_{l_1}(x_1', y_1', t) = \frac{c}{8\pi} \cdot \sum_{l=1}^{3} \tilde{U}_{l_1}(x_1', y_1', t) \cdot \tilde{U}_{l_1}^*(x_1', y_1', t). \qquad (3.109)$$

In view of the smallness of the angles θ and α, we will subsequently neglect the component of the diffracted field along z_0'.

Assuming that the shape of the light pulse corresponds to the Gaussian distribution law, and also, assuming the coherence time of the light source, $\tau_{coh} = \dfrac{l_{coh}}{c}$, where l_{coh} is the longitudinal coherence length, is much smaller than the radiation pulse duration (this is done, for example, for a copper vapor laser) the time-average distribution of the light intensity in the line on the screen:

$$I(x_1', y_1') = \frac{1}{T} \int_{-\infty}^{\infty} e^{-\frac{t^2}{\tau_0^2} \cdot 4\ln 2} \cdot I_1(x_1', y_1', t) dt =$$

$$= \Gamma_0 \left[D_0 + m_0 \cdot e^{-\left(\frac{\pi \cdot f_0 \cdot \tau_0}{2\sqrt{\ln 2}}\right)^2} \cdot \sqrt{D_1^2 + D_2^2} \cdot \sin\left(\frac{2\pi \cdot f_0}{v} \cdot x_1' + \Delta_1\right) + \right. \quad (3.110)$$

$$\left. + \frac{m_0^2}{2} \cdot e^{-\left(\frac{\pi \cdot 2 f_0 \cdot \tau_0}{2\sqrt{\ln 2}}\right)^2} \cdot \sqrt{D_3^2 + D_4^2} \cdot \sin\left(\frac{4\pi \cdot f_0}{v} \cdot x_1' + \Delta_3\right) \right],$$

where the following notations are introduced: T is the period by which time integration takes place, τ_0 is the duration of the light pulse in the level of 0.5 light intensity,

$$D_0 = \sum_{l=1}^{2} \left\{ J_{l0}^2(0) + J_{l1}^2(0) + \frac{m_0^2}{4} \cdot [J_{l0}^2(+1) + J_{l1}^2(+1) + J_{l0}^2(-1) + J_{l1}^2(-1)] \right\}, \quad (3.111)$$

$$D_1 = \sum_{l=1}^{2} \{J_{l0}(0) \cdot [J_{l0}(+1) + J_{l0}(-1)] + J_{l1}(0) \cdot [J_{l1}(+1) + J_{l1}(-1)]\}, \quad (3.112)$$

$$D_2 = \sum_{l=1}^{2} \{J_{l0}(0) \cdot [J_{l1}(-1) - J_{l1}(+1)] + J_{l1}(0) \cdot [J_{l0}(+1) - J_{l0}(-1)]\}, \quad (3.113)$$

$$D_3 = \sum_{l=1}^{2} [J_{l0}(+1) \cdot J_{l0}(-1) + J_{l1}(+1) \cdot J_{l1}(-1)], \quad (3.114)$$

$$D_4 = \sum_{l=1}^{2} [J_{l0}(+1) \cdot J_{l1}(-1) - J_{l1}(+1) \cdot J_{l0}(-1)], \quad (3.115)$$

$$\Delta_1 = \arctg \frac{D_1}{D_2}, \quad \Delta_2 = \arctg \frac{D_3}{D_4}, \quad (3.116)$$

$$J_{l0}(p) = \Upsilon(\beta_{p,0}^{(1)}) \cdot \int_{0}^{\tilde{\alpha}} (-1) \cdot \bar{K}_1(\theta_p^{(1)}, \theta_0^{(0)}, \alpha) \cdot \sin c \left[\frac{L_0}{\pi} \cdot (\gamma_p^{(1)} + \gamma_0^{(0)}) \cdot \alpha^2 \right] \times$$
$$\times \sin[2L_0 \cdot (1 - \gamma_p^{(1)}) \cdot \alpha^2] d\alpha, \quad (3.117)$$

$$J_{11}(p) = \Upsilon(\beta_{p,0}^{(1)}) \cdot \int_0^{\tilde{\alpha}} \overline{K}_1(\theta_p^{(1)}, \theta_0^{(0)}, \alpha) \cdot \sin c\left[\frac{L_0}{\pi} \cdot (\gamma_p^{(1)} + \gamma_0^{(0)}) \cdot \alpha^2\right] \times$$
$$\times \sin[2L_0 \cdot (1 - \gamma_p^{(1)}) \cdot \alpha^2] d\alpha, \tag{3.118}$$

$$J_{20}(p) = \Upsilon(\beta_{p,0}^{(1)}) \cdot \int_0^{\tilde{\alpha}} \overline{K}_2(\theta_p^{(1)}, \theta_0^{(0)}, \alpha) \cdot \sin c\left[\frac{L_0}{\pi} \cdot (\gamma_p^{(1)} + \gamma_0^{(0)}) \cdot \alpha^2\right] \times$$
$$\times \cos[2L_0 \cdot (1 - \gamma_p^{(1)}) \cdot \alpha^2] d\alpha, \tag{3.119}$$

$$J_{21}(p) = \Upsilon(\beta_{p,0}^{(1)}) \cdot \int_0^{\tilde{\alpha}} \overline{K}_2(\theta_p^{(1)}, \theta_0^{(0)}, \alpha) \cdot \sin c\left[\frac{L_0}{\pi} \cdot (\gamma_p^{(1)} + \gamma_0^{(0)}) \cdot \alpha^2\right] \times$$
$$\times \sin[2L_0 \cdot (1 - \gamma_p^{(1)}) \cdot \alpha^2] d\alpha, \tag{3.120}$$

$$\overline{K}_1(\theta_p^{(1)}, \theta_0^{(0)}, \alpha) = A(\theta_p^{(1)}, \theta_0^{(0)}, \alpha) \cdot \sin\left[\frac{2\pi}{\lambda_0} n_0 \cdot \alpha \cdot (y_1' - \alpha_0 \cdot F_1)\right] +$$
$$+ B(\theta_p^{(1)}, \theta_0^{(0)}, \alpha) \cdot \cos\left[\frac{2\pi}{\lambda_0} n_0 \cdot \alpha \cdot (y_1' - \alpha_0 \cdot F_1)\right], \tag{3.121}$$

$$\overline{K}_1(\theta_p^{(1)}, \theta_0^{(0)}, \alpha) = C(\theta_p^{(1)}, \theta_0^{(0)}, \alpha) \cdot \cos\left[\frac{2\pi}{\lambda_0} n_0 \cdot \alpha \cdot (y_1 - \alpha_0 \cdot F_1)\right] +$$
$$+ D(\theta_p^{(1)}, \theta_0^{(0)}, \alpha) \cdot \sin\left[\frac{2\pi}{\lambda_0} n_0 \cdot \alpha \cdot (y_1 - \alpha_0 \cdot F_1)\right], \tag{3.122}$$

$$A(\theta_p^{(1)}, \theta_0^{(0)}, \alpha) = \frac{\alpha}{\sqrt{(1 + \rho_1^2) \cdot (\alpha^2 + \theta_0^2) \cdot (\alpha^2 + \theta_p^2)}} \times$$
$$\times \{(1 + \rho_1 \cdot \rho_0) \cdot (\alpha^2 - \theta_p \cdot \theta_0) + \rho_1 \cdot (\rho_1 + \rho_0) \cdot \theta_p \cdot (\theta_p + \theta_0)\}, \tag{3.123}$$

$$B(\theta_p^{(1)}, \theta_0^{(0)}, \alpha) = \frac{\alpha}{\sqrt{(1 + \rho_1^2) \cdot (\alpha^2 + \theta_0^2) \cdot (\alpha^2 + \theta_p^2)}} \times$$
$$\times \{\rho_1 \cdot \theta_1 \cdot (1 + \rho_1 \cdot \rho_0) \cdot (\alpha^2 - \theta_p \cdot \theta_0) - \alpha^2 \cdot (\rho_1 + \rho_0) \cdot (\theta_p + \theta_0)\}, \tag{3.124}$$

$$C(\theta_p^{(1)}, \theta_0^{(0)}, \alpha) = \frac{\alpha}{\sqrt{(1 + \rho_1^2) \cdot (\alpha^2 + \theta_0^2) \cdot (\alpha^2 + \theta_p^2)}} \times$$
$$\times \{\theta_p \cdot (1 + \rho_1 \cdot \rho_0) \cdot (\alpha^2 - \theta_p \cdot \theta_0) - \rho_1 \cdot (\rho_1 + \rho_0) \cdot \alpha^2 \cdot (\theta_p + \theta_0)\}, \tag{3.125}$$

$$D(\theta_p^{(1)},\theta_0^{(0)},\alpha) = \frac{\alpha}{\sqrt{(1+\rho_p^2)\cdot(\alpha^2+\theta_0^2)\cdot(\alpha^2+\theta_p^2)}} \times$$

$$\times\{\rho_p\cdot(1+\rho_p\cdot\rho_0)\cdot(\alpha^2-\theta_p\cdot\theta_0)-(\rho_p+\rho_0)\cdot\theta_p\cdot(\theta_p+\theta_0)\},$$

(3.126)

$$\Gamma_0 = \frac{c\cdot E_0^2\cdot\chi_0^2\cdot\tau_0\cdot F_1}{4\cdot\sqrt{\pi}\cdot\sqrt{\ln 2}\cdot\lambda_0\cdot M_1'\cdot M_2'\cdot T},$$

(3.127)

$\theta_p = \theta_p^{(1)}$, $\theta_0 = \theta_0^{(0)}$, $\rho_p = \rho\left(\theta_p^{(1)},\alpha\right)$ are determined from the expressions (3.49), and the angles θ_p and θ_0 are found from the system of equations (3.60), E_0 is the electric field strength of the light wave in the air.

The first term in expression (3.110) determines the constant component in the image signal, the second – modulation with the frequency of the input signal, the third – non-linear distortions in the image with twice the frequency of the signal. The coefficient Γ_0 characterizes the change in the mean intensity of light in a line on the screen due to the conversion of the dimensions of the light beam and the increase in the amplitudes of the light and ultrasonic fields.

3.3. Frequency-contrast characteristic and the limiting number of solvable system elements per frame

3.3.1. The case of low efficiency of the acousto-optical interaction

Expression (3.110) is a two-dimensional distribution of the light intensity in the image of line $I(x_1, y_1)$, modulated by a harmonic law by the ultrasonic signal, which is obtained in the first-order approximation of the interaction. This approximation corresponds to the case of low diffraction efficiency. We use this expression to find the frequency–contrast characteristic of the system (FCC). To this end, consider the cross-section of this distribution by the plane $(y = y_0)$, for which $I(y_0) = I_{\max}$. Define the contrast transfer function as a function of the modulation frequency f_0 as

$$M(f_0) = \frac{I_{\max}(f_0)-I_{\min}(f_0)}{I_{\max}(f_0)+I_{\min}(f_0)},$$

(3.128)

which will be the desired FCC of the line imaging system. The function (3.128) was calculated for different values of the duration of the light pulse τ_0, the cone of the angles of incidence of light to the sound beam 2α, which corresponded to the choice of different focal lengths F_1, for the central ultrasound frequencies f and the Bragg angles $\theta_0^{(0)}$, $\theta_0^{(1)}$ corresponding to these frequencies, and also for different lengths of the acousto-optical interaction L. The acousto-optical interaction medium is the TeO_2 crystal. The wavelength of the light emission in the calculations was chosen equal to 510.6 nm, which corresponds to the green emission line of the copper vapor laser, and the depth of modulation of the ultrasonic wave $m_0 = 1$.

Calculations show that $M(f_0)$ practically does not depend on the cone of angles of incidence $2\tilde{\alpha}$ in the plane orthogonal to the scattering plane. Its shape is mainly determined by the choice of the carrier frequency of ultrasound f, the width of the ultrasonic beam L and the duration of the light pulse τ_0.

Figures 3.8–3.10 show the calculated families for the variation of the main parameters of the acousto-optical interaction: the duration of the light pulse τ_0, the interaction length L, and the carrier frequency of ultrasound f.

Fiure. 3.8 shows the dependence of the system of the system on the duration of the light pulse τ_0. Calculations were made with the following parameters: carrier frequency of ultrasound $f = 80$ MHz, interaction length $L = 4$ mm, focal length of the lens $F_1 = 0.5$ m.

Curves *1–5* correspond to the durations of the light pulse $\tau_0 = 5$ ns (*1*), 10 ns (*2*), 20 ns (*3*), 30 ns (*4*), 40 ns (*5*). From Fig. 3.8 it is seen that as the duration of the light pulse decreases, the decay of the FCC is slower.

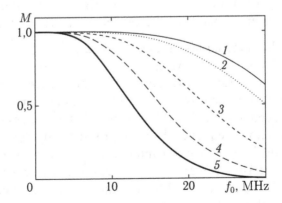

Fig. 3.8. Dependence of the FCC of the system on the duration of the light pulse τ_0.

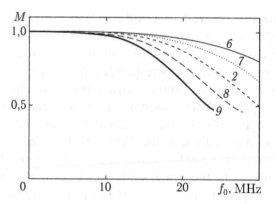

Fig. 3.9. Dependence of the FCC of the system on the interaction length L.

Figure 3.9 shows the FCC of the system at $f = 80$ MHz, $\tau_0 = 10$ ns and different values of the interaction length L. Curves *2, 6–9* correspond to $L = 4$ mm (*2*), 2 mm (*6*), 3 mm (*7*), 5 mm (*8*) and 6 mm (*9*).

As the size of the ultrasonic beam (the interaction length L) decreases, the decay of the FCC occurs more slowly, which is explained by the increase in the spatial–frequency region of the acousto-optical interaction due to the broadening of the directivity diagram of the emitter of elastic waves. In all cases, a decrease in the length L leads to an increase in contrast in the transmission of the image.

Figure 3.10 shows the FCC of the system at $L = 4$ mm, $\tau_0 = 10$ ns, and different carrier frequencies of ultrasound f. Curves *2, 10–14* correspond to $f = 80$ MHz (*2*), 65 MHz (*10*), 70 MHz (*11*), 75 MHz (*12*), 85 MHz (*13*) and 90 MHz (*14*). Figure 3.10 shows that the FCC of the system expands as the carrier frequency of the ultrasound approaches the degeneracy frequency $f = f_B' = \dfrac{v}{\lambda_0} n_0 \sqrt{(n_0^2 + n_e^2) \cdot G_{33}}$. At $\lambda_0 = 510.6$ nm $f_B' = 57.9$ MHz.

In a real modulator, the choice of the carrier frequency of ultrasound is determined by: a) choosing the midpoint at the angular-frequency dependence $\theta_i(f)$, $\theta_d(f)$, at which the necessary acousto-optical interaction frequency band is provided; b) the absence of a dip caused by the two-phonon interaction in the band of operating frequencies of the AOM; c) an acceptable value of ultrasonic attenuation. In connection with this, the optimal carrier frequency of ultrasound should be chosen from the relation

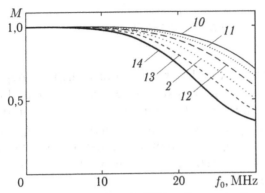

Fig. 3.10. Dependence of the system's FCC on the carrier frequency of ultrasound f.

$$f = f_B' + \frac{\Delta f_0}{2},$$

where Δf_0 is the band of modulating ultrasound frequencies.

Fig. 3.11 shows the dependence of the coefficient

$$N(f_0) = \ln \frac{I_{max}(f_0)}{I_{min}(f_0)} \tag{3.129}$$

from the modulation frequency f_0, constructed for the same functions $I(f_0)$, as in Fig. 3.8.

Since the quantity [154] is taken as the number of luminance grades perceived by the eye,

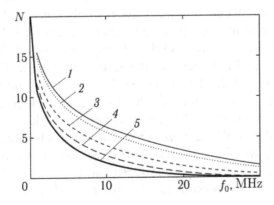

Fig. 3.11. Logarithmic function of contrast transfer.

$$N(f_0) = \frac{1}{\sigma} \cdot \ln \frac{I_{max}}{I_{min}}, \tag{3.130}$$

where $\sigma = 0.02 \div 0.05$ is the threshold contrast, then the dependences in Fig. 3.11 can be useful in estimating the number of luminance gradations (grayscale) and the boundary frequency of the image perceived by the eye. Taking the frequency $f_0 = f_b$ for which $N(f_b) = \sigma$, by solving the corresponding equations, we can find that for $\tau_0 = 40$ ns $f_b \cong 25$ MHz, for $\tau_0 = 30$ ns $f_b \cong 31$ MHz, and at $\tau_0 = 20$ ns $f_b \cong 50$ MHz.

In all cases, an increase in the duration of the light pulse in the range from 5 to 40 ns leads to a decrease in the contrast in the transmission of amplitude-modulated signals. From the graphs in Fig. 3.11 it follows that the theoretical value of the number of luminance gradations at low modulating frequencies, depending on the magnitude of the threshold contrast, can be in the range $200 \div 500$.

In reality, these dependences give excessive values of the function $N(f_0)$ at low and high frequencies, which is due to the non-linear character of the diffraction, which leads to an increase in I_{min} due to the contribution of the modulation frequency components from the higher harmonics. In addition, to I_{min} it is necessary to add noise from the laser associated with the superluminescence of the active substance, from parasitic reflections and scattering on the optical elements of the system, as well as the noise of the speckle structure in the image. On the other hand, when the system is operating, I_{max} has to be reduced in order to work at acceptable values of the nonlinear distortions in the formed image.

It should be noted that when forming an image along a row (coordinate x_1), no special requirements for coherence and spectral purity of laser radiation are presented, since a picture of the image of the ultrasonic field in the AOM is being constructed. The only requirement is the ability to remove the zero and parasitic diffraction orders with the help of diaphragm D. At the same time, the image of the light beam should be formed on the coordinate y_1, therefore high requirements are imposed on the spatial coherence of laser radiation.

To estimate the limiting number of solvable elements of the system by frame, we find the distribution of the light intensity in the generated line along the y_1 coordinate in the image plane. For this we consider the expression (3.110) for $f_0 \simeq 0$, which becomes

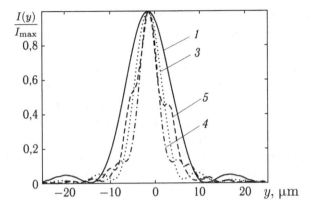

Fig. 3.12. Normalized functions of the transverse distribution of the intensity of diffracted light in the focal plane of the lens with different focal lengths F_1, m: 0.5 (*1*); 0.30 (*3*); 0.2 (*4*); 0.15 (*5*) at $f = 80$ MHz, $L = 4$ mm, $\lambda_0 = 510.6$ nm.

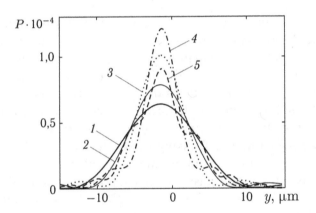

Fig. 3.13. Transverse distributions of the intensities of diffracted light in the focal plane of the lens with different focal lengths F_1, m: 0.5 (*1*); 0.4 (*2*); 0.3 (*3*); 0.2 (*4*); 0.15 (*5*) at $f = 80$ MHz, $L = 4$ mm, $\lambda_0 = 510.6$ nm.

$$I(y_1) = \Gamma_0 \cdot (1 + m_0)^2 \cdot [J_{10}^2(0) + J_{11}^2(0) + J_{20}^2(0) + J_{21}^2(0)]. \qquad (3.131)$$

In Fig. 3.12 are graphs of the normalized functions $I(y)/I_{max}$, and in Fig. 3.13 the dependences $P = I(y)/\Gamma_0 \cdot (1 + m_0)^2$, which are distributions of the relative intensities of the diffracted field along the y coordinate in the focal plane of lens L_1, constructed for different focal lengths F_1 of this lens. All graphs are plotted for the normal angle of incidence of a plane light wave on the input lens $\alpha_0 = 0$. The fall of a light beam on an AOM crystal at an angle $\theta_0^{(0)}$ to the optical axis in the diffraction plane leads to the position of the maxima of the intensity distribution being displaced with respect to the straight line $y = 0$ by 1.5 μm.

The dimensions of the generated image on the screen are $y_1 = y \cdot M_2'$, where M_2' is the magnification factor of the telescopic system L_2, L_3 (see Fig. 3.6), and the intensity $I(y_1) = \Gamma_0 \cdot (1 + m_0)^2 \cdot P$. The graphs, indicated by the numbers *1–5* correspond to the external angles of convergence of the light beam $\alpha_{external}$ 0,02; 0.025; 0.033; 0.05 and 0.067 rad. At $F_1 = 0.5$ m and $\alpha_{external} \ll 0.02$ rad, the transverse distribution of the light intensity in the focal plane of the lens is close to the ideal picture of the diffraction of a plane light wave in the absence of a crystal:

$$I(y)/I_{max} = \sin^2\left(\frac{\pi + a_0 \cdot y}{\lambda_0}\right) \bigg/ \left(\frac{\pi + a_0 \cdot y}{\lambda_0}\right)^2. \tag{3.132}$$

With an increase in the cone of the external angles of incidence of light on the AOM crystal, the side lobes rise and the intensity of the light field intensity distribution gradually increases. Starting at $F_1 = 0.2$ m ($\alpha_{external} \geq 0.05$ rad), there is a noticeable decrease in intensity at the maximum of the transverse distribution, an increase in its width, and a symmetry distortion in the form of a transverse light intensity distribution. Increasing the intensity at the maximum of the image signal due to focusing does not occur. At $F_1 = 0.15$ m ($\alpha_{external} = 0.67$ rad), the maximum intensity value decreases.

This behaviour of the transverse distribution can be explained by the distortions of the wave front in the diffracted light field along the y coordinate that result from the fact that the amplitudes and phases of the diffracted plane waves propagating at different angles α to the optical axis of the crystal are different. These waves acquire unequal phase delays when passing through the region of interaction with ultrasound.

Figure 3.13 shows that the acousto-optical modulator from TeO$_2$ will practically have no appreciable effect on the distribution of the intensity of the light field in the row along the y coordinate and, consequently, on the number of decidable elements of the system along the frame up to angles $\alpha_{external} \simeq 2$ deg, which corresponds to $F_1 = 0.3$ m at $\lambda_0 = 510.6$ nm and a light beam with a diameter of 2 cm.

The real light beam incident on the input lens L_1 is a superposition of plane waves that fall on the lens at different angles α_0 within the angular divergence of the entire beam $\Delta\alpha_0$. In this case, the resulting distribution of the field along the y coordinate in the focal plane of the lens at the output of the AOM will be the sum of the individual distributions whose centers are shifted relative to one another in the range $\Delta y = \Delta\alpha_0 \cdot F_1$.

The actual number of solvable elements per frame will depend on the actual width of the radiation pattern of the laser radiation, which is determined by the range of angles $\Delta\alpha_0$, and on the factors leading to its broadening after the passage of light through the optical forming system. These factors, in addition to large convergence angles, may be inhomogeneities in the refractive index of the optical elements of the system, leading to distortion of the light wave front, as well as unevenness in the distribution of the light field at the output aperture of the deflector, caused, for example, by inhomogeneous illumination of the aperture by incident light [165] or the dissimilar diffraction efficiency due to damping of the elastic wave in the acousto-optical deflector [166].

3.3.2. Influence of the nonlinearity of the acousto-optical interaction on the quality of the image being formed. Approximation of the third order of interaction. Limits of applicability of the first order of interaction

A feature of the AOM operation when forming a television image is the essential non-linearity of its amplitude transfer characteristic, which leads to a decrease in the operating dynamic range of the system and distortion of the light signal at high diffraction efficiencies. To estimate the influence of the non-linearity of the process of light diffraction on ultrasound on the formation of the image of the line, when finding the diffracted field, we take into account the third order of interaction. Assuming that the amplitude of the input ultrasonic signal varies according to the harmonic law with the modulation depth m_0 (3.71), we use expression (3.73) for the field at the output aperture of the acousto-optical modulator. In connection with the fact that, as we found, in the first-order approximation, the contrast is practically independent of the angles α, we assume that the focal length of the input cylindrical lens is sufficiently large. In this case, the dependence of the amplitudes and phases of the spectral components of the diffracted field on the angle α can be neglected. In the same way as we did in Sec. 2.2, for each Cartesian component of the field (3.73) we find the instantaneous distribution of the light intensity in the screen plane along the coordinate x_1. Summarizing the contributions from all the components of the field, and also by averaging over time, similarly to (3.110), we find the time-average distribution of the light intensity in the line on the screen:

$$I_1(x_1) = \widetilde{\Gamma}_0 \cdot \chi_0^2 \left\{ C_0 + 2 \cdot \sum_{n=1}^{6} \left[C_{n1} \cdot \cos\left(\frac{2\pi \cdot n \cdot f_0 \cdot x_1}{v}\right) + \right. \right.$$

$$\left. \left. + C_{n2} \cdot \sin\left(\frac{2\pi \cdot n \cdot f_0 \cdot x_1}{v}\right) \right] \cdot e^{-\left(\frac{\pi \cdot n \cdot f_0 \cdot \tau_0}{2\sqrt{\ln 2}}\right)^2} \right\},$$
(3.133)

where

$$C_0 = \sum_{k=1}^{2} \cdot \sum_{l=-3}^{3} (A_{lk}^2 + B_{lk}^2),$$
(3.134)

$$C_{n1} = \sum_{k=1}^{2} \cdot \sum_{l=-3}^{3} \cdot \sum_{m=-3}^{3} (A_{lk} \cdot A_{mk} + B_{lk} \cdot B_{mk}) \cdot \delta(l-m-n),$$
(3.135)

$$C_{n2} = \sum_{k=1}^{2} \cdot \sum_{l=-3}^{3} \cdot \sum_{m=-3}^{3} (A_{lk} \cdot B_{mk} + B_{lk} \cdot A_{mk}) \cdot \delta(l-m-n),$$
(3.136)

$$A_{l1} = \cos\varphi(\theta_l) \cdot D_l(\theta_l) - \rho(\theta_l) \cdot \sin\varphi(\theta_l) \cdot H_l(\theta_l),$$
$$B_{l1} = \cos\varphi(\theta_l) \cdot H_l(\theta_l) + \rho(\theta_l) \cdot \sin\varphi(\theta_l) \cdot D_l(\theta_l),$$
$$A_{l2} = \sin\varphi(\theta_l) \cdot D_l(\theta_l) + \rho(\theta_l) \cdot \cos\varphi(\theta_l) \cdot H_l(\theta_l), \qquad (3.137)$$
$$B_{l2} = \sin\varphi(\theta_l) \cdot H_l(\theta_l) - \rho(\theta_l) \cdot \cos\varphi(\theta_l) \cdot D_l(\theta_l),$$
$$D_l = \mathrm{Re}(\tilde{S}_l), \quad H_l = \mathrm{Im}(\tilde{S}_l),$$

$$\tilde{\Gamma}_0 = \frac{c \cdot \tau_0 \cdot a_0^2 \cdot E_0^2}{16 \cdot \lambda_0 \sqrt{\pi \cdot \ln 2} \cdot M_1' \cdot M_2' \cdot T \cdot F_1},$$
(3.138)

$$\tilde{S}_0 = K_{12}(\theta_0^{(1)}, \theta_0^{(0)}) - \frac{\chi_0^2}{8} \left\{ S(0,0,0) + \frac{m_0^2}{4} \cdot [S(0,\ 0,\ \pm 1) + \right.$$

$$\left. + S(0, \pm 1, \pm 1) + S(0, \pm 1, 0)] \right\},$$
(3.139)

$$\tilde{S}_{\pm 1} = \frac{\mathrm{m}_0}{2} \{ K_{12}(\theta_{\pm 1}^{(1)}, \theta_0^{(0)}) \cdot \Upsilon(\beta_{\pm 1,0}^{(1)}) -$$

$$-\frac{\chi_0^2}{8} \Big[S(\pm 1, \pm 1, \pm 1) + S(\pm 1, \pm 1, 0) + \tag{3.140}$$

$$S(\pm 1, 0, 0) + \frac{\mathrm{m}_0^2}{4} \cdot [S(\pm 1, 0, \pm 1) + S(\pm 1, 0, \mp 1) + S(\pm 1, \pm 2, \pm 1)] \Big] \Big\},$$

$$\tilde{S}_{\pm 2} = \frac{\chi_0^2}{8} \cdot \frac{\mathrm{m}_0^2}{4} \cdot [S(\pm 2, \pm 2, \pm 1) + S(\pm 2, \pm 1, \pm 1) + S(\pm 2, \pm 1, 0)], \tag{3.141}$$

$$\tilde{S}_{\pm 3} = \frac{\chi_0^2}{8} \cdot \frac{\mathrm{m}_0^3}{8} \cdot S(\pm 3, \pm 2, \pm 1), \tag{3.142}$$

$S(p, m, n)$ are determined from the expressions (3.78)÷(3.81).

It is seen from expression (3.133) that when the third order of interaction is taken into account, in addition to the components C_{11} and C_{12} describing the modulation with the fundamental frequency f_0, the components of the image signal will contain components with higher harmonics, up to the 6th harmonic of the input signal, leading to its distortion. For a numerical estimate of the degree of distortion, we introduce the coefficient η_{nel} characterizing the root-mean-square deviation of the intensity averaged over the period of the fundamental frequency in the image signal from the component with the frequency of the first harmonic f_0:

$$\eta_{\mathrm{nel}} = \frac{\sqrt{\dfrac{1}{x_m} \cdot \displaystyle\int_0^{x_m} [I_1(x_1) - I_1(x_1)|_{n=1}]^2\, dx_1}}{|I_1(x_1)|_{n=1} - C_0|}$$

$$= \frac{\sqrt{\displaystyle\sum_{n=2}^{6} e^{-(\phi_0 \cdot n)^2} \cdot (C_{n1}^2 + C_{n2}^2)}}{\sqrt{2} \cdot e^{-(\phi_0)^2} \cdot \sqrt{C_{11}^2 + C_{12}^2}}, \tag{3.143}$$

where $\phi_0 = \dfrac{\pi \cdot f_0 \cdot \tau_0}{2 \cdot \sqrt{\ln 2}}$, $x_m = \dfrac{\upsilon}{f_0}$.

The coefficient η_{nel} was calculated for different values of the modulation index χ_0, the frequency f_0, and the duration of the light pulse τ_0. Figure 3.14 shows the dependence of η_{nel} on the value of χ_0

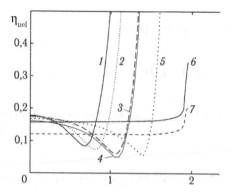

Fig. 3.14. Dependence of the non-linear distortion coefficient η_{nel} from the modulation index χ_0 for different modulation frequencies f_0 MHz at $\tau_0 = 10$ ns: 0.5 MHz (curve *1*), 3 MHz (*2*), 5 MHz (*3*), 7 MHz (*5*), 10 MHz (*6*); at $\tau_0 = 30$ ns: 5 MHz (*4*), 10 MHz (*7*).

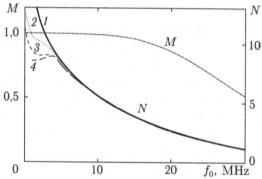

Fig. 3.15. Dependence of contrast transfer functions on different values of the modulation index χ_0: $\chi_0 = 0.01$ curve (*1*); 0.2 (*2*); 0.4 (*3*); 0.8 (*4*).

for various modulating frequencies f_0. Calculations were performed at $m_0 = 1$, $\lambda_0 = 510.6$ nm, $f = 80$ MHz, $L = 4$ mm.

It is seen from the graphs that the non-linear distortions of the harmonic signal in the image with increasing modulation index are different for different modulating frequencies f_0. As the modulation index χ_0 increases, starting with a certain limiting value, which increases with increasing frequency f_0, the non-linear distortions in the image signal increase sharply. At low modulation frequencies, these distortions are particularly large. The increase in the duration of the light pulse leads to a certain decrease in the non-linear distortions with a simultaneous decrease in the contrast in the image.

Fig. 3.15 shows the behavior of the contrast transfer functions $M(f_0)$ and $N(f_0)$, which are analogous to the dependences (3.128)

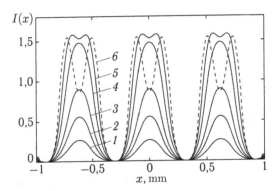

Fig. 3.16. Dependence of the amplitude and shape of the harmonic signal in the image at the modulation frequency $f_0 = 1$ MHz for different values of the modulation index χ_0: 0.2 (*1*); 0.3 (*2*); 0.4 (*3*); 0.6 (*4*); 0.8 (*5*); 1.0 (*6*).

and (3.129), where the absolute values of the intensity maximum and minimum in the image signal are taken under I_{max} and I_{min}. On these graphs, the curves with indices *1, 2, 3, 4* correspond to $\chi_0 =$ 0.01; 0.2; 0.4; 0.8.

It can be seen from the graphs that when the modulation index χ_0 changes, the system of the system is practically unchanged, while the number of possible luminance gradations at low frequencies decreases. The numerical data presented in Fig. 3.15 allow us to calculate the theoretical value of the number of luminance gradations. Using expression (3.123) at $\sigma = 0.02$, we obtain the value $N \simeq 500$ at modulation frequencies of 0.2÷0.5 MHz and $\chi_0 = 0.6÷0.8$.

Figure 3.16 shows the change in the amplitude and shape of the image signal at the modulation frequency $f_0 = 1$ MHz ($f = 80$ MHz, $\tau_0 = 10$ ns, $L = 4$ mm) as the modulation index increases. On these graphs, the distance x, mm ($M_1 = 1$) is plotted along the abscissa axis, and $I(x) = \dfrac{I_1(x_1)}{\tilde{\Gamma}_0}$ along the ordinate axis in accordance with (3.133).

Figure 3.17 shows graphs of the dependence of η_{nel} on the frequency f_0, constructed for different modulation indices: $\chi_{01} = 0.2$ (curve *1*); $\chi_{02} = 0.6$ (*2*); $\chi_{03} = 1.0$ (*3*); $\chi_{04} = 1.2$ (*4*); $\chi_{05} = 1.4$ (*5*) for a fixed value of $L = 4$ mm, $\tau_0 = 10$ ns, $f = 80$ MHz. It can be seen from the graphs that as the modulation index increases the non-linear distortions capture an ever larger region of the operating frequencies.

One way to reduce these distortions is to increase the length of the acousto-optical interaction L. Figure 3.18 shows the calculated

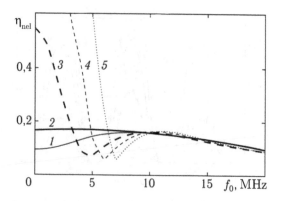

Fig. 3.17. The dependence of the coefficient of non-linear distortions η_{nel} on the frequency f_0 for different values of the modulation index χ_0.

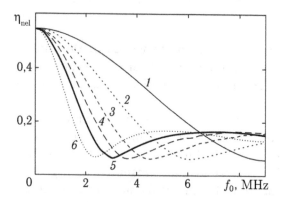

Fig. 3.18. Dependences of the coefficient of non-linear distortions η_{nel} on the frequency f_0 for a fixed value of the modulation index $\chi_0 = 1$ and different lengths L, mm: 2 mm (curve *1*); 3 (*2*); 4 (*3*); 5 (*4*); 6 (*5*); 8 (*6*).

dependences of η_{nel} on the frequency for a fixed value of the modulation index $\chi_0 = 1$ and different lengths L.

The graphs indicated in the figure by numbers *1–6* correspond to lengths L, equal to 2, 3, 4, 5, 6 and 8 mm. The above graphs show that in order to reduce the non-linear distortions, it is necessary to strive to increase the length of the acousto-optical interaction. The length L should be chosen to the maximum possible for a given band of operating frequencies, without significantly degrading the contrast in the image.

Figures 3.19–3.21 show the calculated dependences

$$I(\chi_0) = \frac{I^{+1}_{max}}{I^{+1}_{max} + I^{+0}_{min}},$$ which determine the diffraction efficiency at the

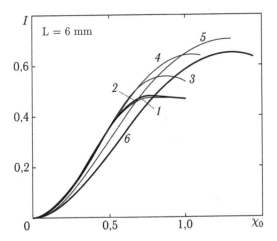

Fig. 3.19. Dependence of the diffraction efficiency in the image $I(\chi_0)$ on the modulation index χ_0 at $L = 6$ mm, $\lambda_0 = 510.6$ nm, $f = 80$ MHz; $\tau_0 = 10$ ns. Curves *1–6* correspond to modulation frequencies of 0.5 MHz (curve *1*), 1 MHz (*2*), 3 MHz (*3*), 5 MHz (*4*), 10 MHz (*5*), 15 MHz (*6*).

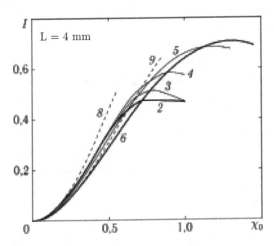

Fig. 3.20. Amplitude transfer characteristics of the system in the third order of interaction for $L = 4$ mm. Curves *8* and *9* correspond to the case of low diffraction efficiency at $f_0 = 1$ and 15 MHz.

maxima of the image signal for a harmonic signal with frequency f_0 from the modulation index.

In these dependences, I^{+1}_{max}, I^0_{min} are the time-average light intensities at the maximum and minimum of the image signal for the +1-st and 0-th diffraction orders; moreover, I^{+1}_{max} was calculated in the third order approximation and corresponds to (3.133), and I^0_{min}– in

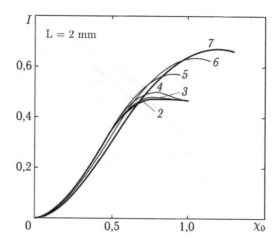

Fig. 3.21. Amplitude transfer characteristics of the system in the third order of interaction for $L = 2$ mm.

the 4th order approximation of the interaction. The expression for $I_0^{(4)}$ (x), the time-average light intensity in the zeroth order of diffraction in the 4th order approximation of the interaction for the harmonic-modulated signal, is presented in Appendix 2.

These dependences can be called the amplitude transfer characteristics of the line imaging system. Calculations of $I(\chi_0)$ were performed for $\lambda_0 = 510.6$ nm, $f = 80$ MHz, $m_0 = 1$, $\tau_0 = 10$ ns for different frequencies f_0 and length L. Figures 3.19–3.21 correspond to $L = 6, 4, 2$ mm. The numbers in the notation of the graphs (*1, 2, 3, 4, 5, 6, 7*) correspond to the modulation frequencies $f_0 = 0.5, 1, 3, 5, 10, 15, 20$ MHz.

Calculations show that the $I(\chi_0)$ dependences for the selected carrier frequency of ultrasound f practically do not change when the modulation frequency f_0 varies from 0.1 to 1 MHz. They also slightly depend on the duration of the light pulse when it varies from 5 to 40 ns.

With a further increase in the modulation frequency for fixed values of L and τ_0, the curves for different f_0 begin to differ substantially from each other. At small indices χ_0, the $I(\chi_0)$ dependences vary in accordance with the quadratic law, which corresponds to the previously considered approximation of low diffraction efficiency. As a criterion for the difference between the two graphs for small modulation indices, it is possible to choose the deviation $\Delta I = \dfrac{I_1^{(1)} - I_1^{(3)}}{I_1^{(3)}} = 0.05$ (5%), where $I_1^{(1)}$ and $I_1^{(3)}$ are the

Table 3.1. The calculated values of χ_{0b} and the diffraction efficiency $I^{(1)}(\chi_{0b})$ corresponding to the deviation $\Delta I_1 = 5\%$

f_0, MHz	χ_{0b}			$I^{(1)}$, o. e.???		
	$L = 2$ mm	$L = 4$ mm	$L = 6$ mm	$L = 2$ mm	$L = 4$ mm	$L = 6$ mm
1	0.20	0.20	0.20	0.075	0.075	0.075
3	0.20	0.23	0.26	0.075	0.100	0.124
5	0.21	0.27	0.30	0.080	0.130	0.150
10	0.26	0.33	0.35	0.120	0.167	0.166
15	0.30	0.35	0.36	0.140	0.160	0.135

diffraction efficiencies calculated in the first-order and third-order approximation, and find χ_{0b} and $I_1^{(1)}(\chi_{0b})$ corresponding to these conditions.

The results of such a comparison are given in Table 3.1 and reflect the applicability limits of the approximation of low diffraction efficiency with a change in the modulation frequency f_0 and the length of the acousto-optical interaction L.

Thus, it can be concluded that the approximation of low diffraction efficiency is limited from above by the diffraction efficiency level of $0.1 \div 0.17$ and the modulation indices χ_{0b} from 0.2 for $f_0 = 1$ MHz to 0.35 for $f_0 = 15$ MHz. With a further increase in the frequency range, the modulation index corresponding to the approximation of low diffraction efficiency slightly increases.

3.3.3. Approximation of the fifth order of interaction. Limits of the applicability of the third-order approximation of the interaction

In order to determine the limits of the applicability of the third-order approximation of the interaction in calculating the picture of the distribution of light in the image, it is necessary to compare the distributions of intensities taking into account the third and fifth orders of interaction. The expression for $I_{+1}^{(5)}(x)$, the time-average light intensity in the image for a harmonic-modulated signal in the approximation of the 5th order of interaction, is presented in Appendix 2. The field was calculated according to the scattering diagram shown in Fig. 3.22. The amplitude transfer characteristics of the acousto-optical imaging system were calculated: $I(\chi_0) = \dfrac{I_{max}^{(+1)}}{I_{max}^{(+1)} + I_{min}^{(0)}}$,

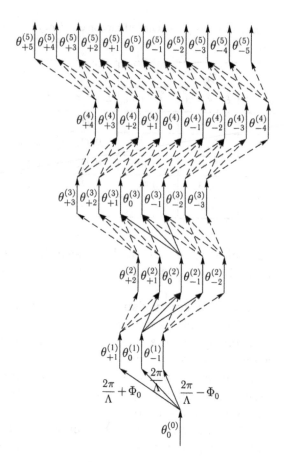

Fig. 3.22. Diagram of scattering for a field in the approximation of the 5th order of interaction.

where $I_{max}^{(+1)}$ was calculated for the third and fifth orders of interaction, and $I_{min}^{(0)}$ for the fourth. Calculations of $I(\chi_0)$ were performed for $\lambda_0 = 510.6$ nm, $f = 80$ MHz, $m_0 = 1$, $\tau_0 = 10$ ns at different frequencies f_0 and lengths $L = 2$ mm, 4 mm and 6 mm.

Figure 3.23 shows the results of calculations for a length $L = 4$ mm. By setting the permissible deviation value $\Delta I \simeq 2\%$, we can determine the boundary values of the modulation index χ_{0gr} and the value $I(\chi_{0b})$ corresponding to this index. The results of the calculations are given in Table 3.2.

Figures 3.24 and 3.25 show the results of calculating the dependence of the diffraction efficiency at the maximum of the image signal $I(\chi_0) = \dfrac{I_{max}^{(+1)}}{I_{max}^{(+1)} + I_{min}^{(0)}}$ on the frequency of the modulating

Table 3.2. The calculated values of χ_{0b} and the diffraction efficiency $I^{(3)}(\chi_{0b})$, corresponding to the deviation $\Delta I_2 = 2\%$

f_0, MHz	χ_{0b}			$I^{(3)}$, o. e.		
	$L = 2$ mm	$L = 4$ mm	$L = 6$ mm	$L = 2$ mm	$L = 4$ mm	$L = 6$ mm
1	0.60	0.64	0.65	0.43	0.46	0.46
3	0.64	0.70	0.75	0.46	0.50	0.54
5	0.65	0.80	0.95	0.47	0.57	0.64
10	0.80	1.05	1.10	0.56	0.68	0.67
15	0.95	1.17	1.15	0.63	0.69	0.62

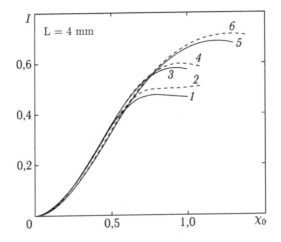

Fig. 3.23. Dependence of the diffraction efficiency in the image $I(\chi_0)$ on the modulation index χ_0 in the third (curves *1, 3, 5*) and the fifth (curves *2, 4, 6*) interaction orders at $L = 4$ mm; for modulation frequencies f_0, MHz: 1 MHz (curves *1, 2*); 5 MHz (*3, 4*) and 10 MHz (*5, 6*).

signal f_0 for different values of the modulation index χ_{0b} and the length of the acousto-optical interaction L. I_{max}^{+1} was calculated in the fifth order approximation of the interaction, and $I_{min}^{(0)}$ – in the fourth approximation. The dependences shown in these figures can be considered as the amplitude–frequency characteristics of the acousto-optical imaging system with a pulsed laser and the AOM made from TeO$_2$.

It can be seen from the graphs that a decrease in the length L leads to an increase in the diffraction efficiency in the high-frequency region and to the broadening of the frequency band of the acousto-optical interaction. An increase in L, on the one hand, increases the diffraction efficiency in the low-frequency region, on the other, reduces the range of operating frequencies of the system. At large

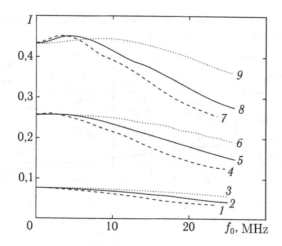

Fig. 3.24. Dependence of the diffraction efficiency in the image $I(\chi_0)$ in the fifth-order approximation of the interaction on the frequency of the modulating signal f_0 for $L = 6,4,2$ mm (dashed, solid and dashed lines, respectively) and for different modulation indices: $\chi_0 = 0.2$ (curves *1, 2, 3*), $\chi_0 = 0.4$ (*4, 5, 6*), $\chi_0 = 0.6$ (*7, 8, 9*).

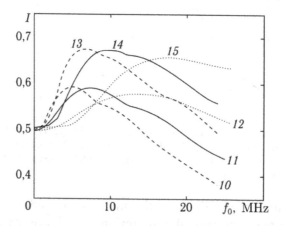

Fig. 3.25. Dependence of the diffraction efficiency in the image $I(\chi_0)$ in the fifth-order approximation of the interaction on the frequency of the modulating signal f_0 for $L = 6,4,2$ mm (dashed, solid and dashed lines, respectively) and for different modulation indices: $\chi_0 = 0.8$ (curves *10, 11, 12*), $\chi_0 = 1.0$ (*13, 14, 15*).

modulation indices, the diffraction efficiency at low frequencies (0.1÷1 MHz) is equalized, which is due to the limitation of the amplitude as a result of the transfer of light energy into the high harmonics of the modulating signal. This efficiency is limited to approximately 0.5 at modulation frequencies up to 1 MHz.

Figure 3.26 shows the dependence of the amplitude transfer characteristics of the system, calculated for the fifth order of

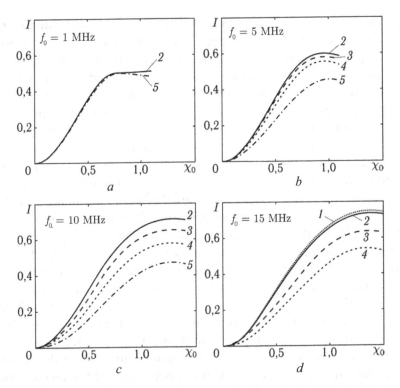

Fig. 3.26. Dependence of the amplitude transfer characteristics of the system in the fifth order of interaction on the duration of the light pulse.

interaction, at the frequencies of the modulating signal 1 MHz (*a*), 5 MHz (*b*), 10 MHz (*c*), 15 MHz (*d*) from the durations of the light pulse. The graphs are plotted for $L = 4$ mm. Curves *1, 2, 3, 4, 5* correspond to the duration of the light pulse $\tau_0 = 5, 10, 30, 50$ and 100 ns.

It can be seen from these graphs that it is quite sufficient to use lasers with a pulse duration $\tau_0 \simeq 30 \div 40$ ns to form an image of an amplitude-modulated ultrasonic signal with a bandwidth of up to $5 \div 7$ MHz (standard TV signal). With the widening of the frequency band of the video signal at high light pulse lengths, the diffraction efficiency at high frequencies drops sharply. Simultaneously, the contrast of these spectral components decreases. The increase in intensity occurs only due to the increase in the signal carrier frequency.

Analyzing the expressions (3.110), (3.133) and the graphs in Figs. 3.14–3.26 in combination with the previously considered dependences, it can be concluded that the degree of distortion of the harmonic signal in the image depends on the orientation conditions of the spectral components of the diffracted light field relative to the center of the acousto-optical interaction band. At the central setting (Bragg interaction conditions are fulfilled for the central spectral component of the amplitude-modulated signal) for small modulation frequencies, the lateral spectral components are located close to the carrier frequency of ultrasound. For them, conditions for repeated diffraction are easier to perform, and, consequently, for the appearance of lateral spectral components at multiple modulation frequencies. In this case, the shape of the harmonic image signal is distorted, it approaches a rectangular one. The increase in intensity at the maximum decreases sharply, since the energy of the light field is efficiently pumped into diffraction orders with multiple modulation frequencies. For higher frequencies f_0, the spectral components in the diffracted light field corresponding to the repeated diffraction, because of the restriction of the acousto-optical interaction band, have a much smaller amplitude than at low frequencies. In them at diffraction, a smaller percentage of the light field energy is pumped. The shape of the signal is distorted to a lesser degree and an increase in the light intensity at the maximum of the image signal is observed with increasing amplitude of the ultrasonic signal. At high frequencies f_0, the decrease in $I(\chi_0)$ is due to the finiteness of the acousto-optical interaction band, and the energy transfer to the lateral spectral components decreases, which leads to a decrease in the contrast in the image and ultimately to a complete loss of information on the modulation of the ultrasonic signal. In this case, the image on the screen is a uniform light exposure with an intensity corresponding to the amplitude of the central spectral component of the signal.

In this chapter, we consider the diffraction of light by an ultrasonic wave with its amplitude modulation at a single frequency. When the ultrasonic carrier frequency is modulated by an input signal with a complex spectrum, consideration of nonlinear distortions in the image signal at large ultrasound amplitudes becomes significantly more complicated. Light fields as a result of diffraction by sound in the first order will contain frequency components shifted in frequency in accordance with the spectrum of the modulating signal. As a result of repeated diffraction, for each frequency component, multiple frequencies appear in accordance with the order of acousto-optical interaction, the amplitudes of which are nonlinearly dependent on the modulation index. Further in the image plane, all the fields are added up and form a pattern of the intensity distribution, in which additional harmonics arise for the fundamental frequencies of the spectrum, as well as the combination components, due to the multiplication of the fields. The increase in the intensity of light in Raman spectral components of a higher order than the first will be determined by the processes of repeated diffraction and occur simultaneously with the transfer of energy into the spectral components with multiple harmonics of the modulating frequencies of the input signal. Therefore, as a result of calculations for one modulating frequency, the values of χ_0 for the modulation indices, from which the nonlinear distortions in the image signal increase sharply, will also hold for a signal with a complex spectrum. To obtain small distortions in the image signal, it is apparently necessary that the modulation indices for all the spectral components of the input modulating signal do not exceed the permissible value χ_0 for its frequency.

In concluding this chapter, it is necessary to clarify the effect of the gyrotropy of the paratellurite crystal on the efficiency of acousto-optical diffraction. In which cases in calculating the amplitude characteristics of deflectors and modulators on paratellurite and other gyrotropic crystals, should it be taken into account? What is the magnitude of this effect and how much does the calculation of gyrotropy change the calculated values of diffraction efficiency?

Figures 3.27 *a*, *b* and 3.28 sheow the calculated dependencies of the shape of the harmonic signal in the image at the modulation frequencies of 1 MHz and 10 MHz from the modulation index are given, taking into account the dependence of the ellipticity of the diffracted waves on the angles $\rho = \rho(\theta)$ and in the case when the ellipticity of all waves is the same and is equal to 1 or zero. In the first case, the calculation results will differ by approximately 20%, in the second by 50%.

It is also interesting to compare the results of calculating the intensity of diffracted light in the limiting Bragg regime with and without taking into account the gyrotropy of the crystal.

The deflectors, as a rule, operate under the Bragg diffraction regime within a small deviation of the frequency along the length of the sound line. In the case of diffraction at one frequency, which fills the deflector aperture, the repeated diffraction occurs with the same scattering coefficients. The light intensity in the +1-th diffraction order is given by expression

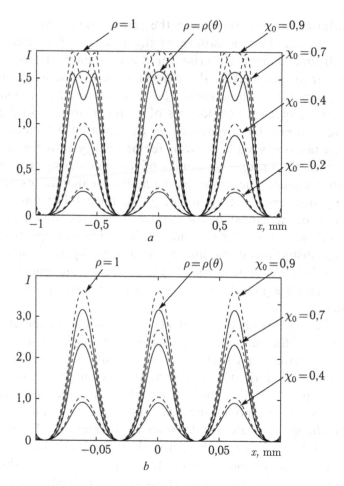

Fig. 3.27. Dependence of the shape of the harmonic signal in the image at $f_0 = 1$ MHz (*a*) and $f_0 = 10$ MHz (*b*) on the modulation index χ_0 ($L = 4$ mm), taking into account the gyrotropy $\rho = \rho(\theta)$ and $\rho = 1$ (circular polarization light waves).

$$\frac{I_{+1}}{I_0} = K \cdot \left(\sum_{n=0}^{\infty} \frac{(-1)^n \cdot \chi_0^{(2n+1)}}{(2n+1)!} \cdot K^{(n)} \right)^2, \tag{3.144}$$

where $K = \dfrac{\left(1 + \rho_0 \cdot \rho_{+1}\right)^2}{\left(1 + \rho_{+1}^2\right)\left(1 + \rho_0^2\right)} \approx 1$.

In this case, $\dfrac{I+1}{I_0} \approx \sin^2(\chi_0)$ and the correction to the intensity of the diffracted light due to gyrotropy is small.

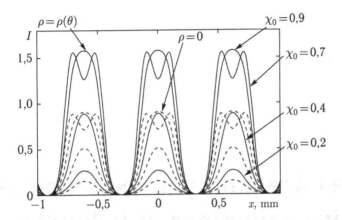

Fig. 3.28. Dependence of the shape of the harmonic signal in the image at $f_0 = 1$ MHz on the modulation index χ_0 ($L = 4$ mm), taking into account the gyrotropy $\rho = \rho(\theta)$ and at $\rho = 0$ (linear polarization of light waves).

When an acousto-optical device is used as a modulator, diffraction takes place simultaneously on several components of the signal spectrum, for which corrections that take into account gyrotropy (scattering coefficients) are different. In this case, the deviations from the calculations taking into account the gyrotropy increase with the modulation index and the band of modulating frequencies.

4

Development and experimental investigations of individual sections and elements of acousto-optical system for the formation of TV images

4.1. Optimization of the parameters of the copper vapour laser radiation applied to the projection information display system

Theoretical consideration of the process of forming a line image in an acousto-optical system with a pulsed laser shows that the quality of the reproduced information should be significantly influenced by such parameters of light radiation as the duration of the generation pulse and the divergence of laser radiation. The long duration of the generation pulse should lead to a decrease in the limiting number of solvable elements and a decrease in the contrast in the image of the amplitude-modulated signals. The divergence of the laser radiation should correspond to the capabilities of the deflector used for the light emission along the frame. If an acousto-optical deflector is used in the system, it should be close to diffraction. In addition, for effective use of light power, the polarization of the laser radiation incident on the acousto-optical modulator must correspond to the eigenmode of the light wave propagating in the interaction medium (3.45)–(3.47). In addition to the requirements for radiation parameters, the laser itself must operate reliably and efficiently at frequencies corresponding to the standard horizontal pulse repetition frequency, for example, 15.625 kHz for the SECAM system, and also have an external trigger from these line pulses. The development of a laser that meets all of the above requirements is the first task that must be solved when creating an information display device whose operation principle is based on the impulse method of forming a line. To solve this problem, the possibility of optimizing the parameters of a copper vapour laser was studied on the basis of the sealed-off gas-discharge active elements (Kulon, Kulon-CM, Kristall, etc.), produced by the domestic industry, with reference to the television display system information.

At the beginning of the work on the projected pulsed laser system (1983–1985), solid-state pulsed lasers with lamp pumping and with the conversion of radiation to the second harmonic in many of their parameters did not satisfy the above requirements, and lasers on copper and gold vapour with sealed-off active elements of domestic production (Istok Company, Fryazino, Moscow Region), although they were available, but had bulky and unreliable power supplies that did not allow demonstration of the system's operation beyond the limits of the laboratories. In addition, the output parameters of the radiation of these lasers did not adequately meet the requirements formulated above. In this regard, great efforts were made to develop small-sized power supplies and a system for pumping copper vapour lasers for a mobile projection laser system.

Generation of a light pulse in a copper vapour laser takes place at the front of a pulse of electric power supplied to the active element, and experiments show that the generation parameters are significantly improved with a decrease in the excitation pulse [70, 71]. As electrical commutators, hydrogen thyratrons are used in most practical schemes for pulse pumping of metal vapour lasers, the traditional circuit for incorporating them into the discharge circuit with the active element is shown in Fig. 4.1 *a*. In the process of switching the electrical energy stored in the capacitor C_0, its redistribution occurs between the active element (AE) of the laser and thyratron T_1, depending on their conductivity. The sharpening capacitance C_1 serves to compensate for the equivalent inductance of the discharge channel of the active element.

The final time of establishing the conductivity in the thyratron results in large switching losses in it when short pump pulses are formed, which reduces the life of the thyratron and the reliability of the entire device. The time for establishing the small conductivity of

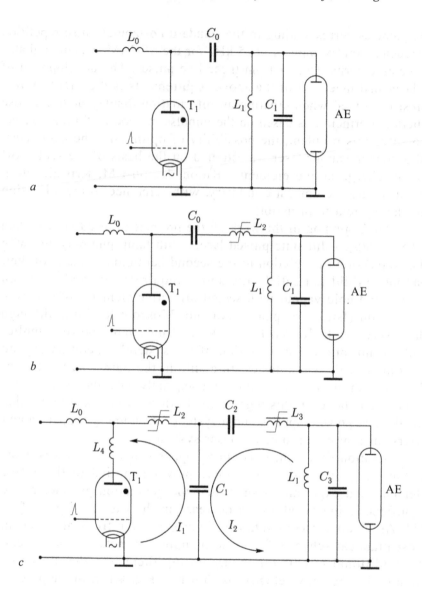

Fig. 4.1. Electrical circuits for switching the working capacitance to the active element of a copper vapour laser.

the thyratron determines also the duration of the pump pulse applied to the active element of the laser. The reliability of such a circuit at high pulse repetition rates is especially reduced.

In Refs. [71, 167], the problem of reducing commutation losses in a thyratron was solved by dividing the discharge and switching circuits through a step-up autotransformer. In the discharge circuit, a magnetic compression link was used, which made it possible to increase the rate of increase of the power pulse on the active element at sufficiently small commutation losses on the thyratron. As the authors of these papers report, the scheme improved in this way allowed to increase the output power of laser radiation at a pulse repetition rate of 8 kHz in comparison with the circuit in Fig. 4.1 *a*. Nevertheless, an attempt to implement a similar pump scheme for the repetition rate pulses of 15.625 kHz with an average power taken from the rectifier $2.5 \div 3$ kW, showed that, despite the decrease in the duration of the pump pulse, the efficiency of the device is substantially reduced due to large thermal losses in the magnetic compression link. In order to provide the necessary average power supplied to the active laser element, it is necessary to increase the switched energy, which leads to additional losses on the thyratron.

A significant increase in the pumping efficiency was achieved in the circuit of the discharge circuit shown in Fig. 4.1 *b*. In this circuit, a saturated choke L_2 is introduced into the discharge circuit of the thyratron T_1, which carries out a time delay $t_0 = 20 \div 50$ ns between the start pulse and the start of the discharge of the working capacitance C_0 through the thyratron to the active element. Since the discharge of the capacitance is carried out with a high conductivity of the thyratron, the switching losses on it are substantially reduced. Experiments carried out with the active element UL-101 and thyratron TGI-2000/35 showed that with an operating capacitance of $C_0 = 1500$ pF, $C_1 = 430$ pF with an average power taken from the rectifier 2.5 kW, the average power generation at the repetition rate of 15.625 kHz pulses increases approximately by two times in comparison with the circuit in Fig. 4.1 *a* [182]. A similar increase in the output power of generation was also obtained with the active elements TLG-5 and GL-201 and, respectively, with thyratrons TGI-1000/25 and TGI-2000/35 at a frequency of 15.625 kHz.

Observation of the oscillograms of the current pulses through the active element and the voltage applied to it showed that when this non-linear choke is introduced into the discharge circuit, the current pulse front decreases from 100 to 50 ns for the AE UL-101 and the thyratron TGI-2000/35, the delay between the current and voltage pulses on the active element decreases, which leads to a decrease in the duration and increase in the peak electric power of the pump pulses. This, apparently, explains the increase in the average output power of laser radiation. Measurement of the duration of the light pulse for a laser with an active element of UL-101 showed that it decreased by approximately three times as compared with the scheme without a saturable choke and amounted to 10 ns at the half-power level. The shape of the radiation pulse approximately corresponded to the Gaussian distribution law.

When working with more powerful sealed-off active elements (GL-201, GL-205A) at a repetition frequency of 15.625 kHz, an even larger increase in the pumping efficiency was achieved with the circuit of the discharge circuit shown in Fig. 4.1 c.

This scheme works on the principle of doubling the voltage on the active element (AE). For this, the working capacitance C_0 is divided into two capacitances C_1 and C_2 by the $C_0/2$, such that two discharge circuits are formed. One circuit includes a thyratron T_1, a capacitance C_1, a constant inductance L_4 of $0.6 \div 1$ μH and a non-linear inductance of L_2, and the second is an AE, two C_1 and C_2 capacitors connected in series, and a saturating reactor L_3. When a triggering pulse is applied to the thyratron, one of the capacitors C_1 is recharged through the thyratron, the inductance L_2, whose core is in the saturated state, and the discharge inductance L_4. Inductance L_4 limits the pulse current I_1 and reduces the switching losses in the thyratron. The non-linear choke L_3 produces delays between the discharges of the two capacitances C_1 and C_2. The value of the delay is chosen such that the beginning of the discharge of capacitance C_2 corresponds to the moment when the first capacitance C_1 has time to be recharged to a negative voltage. In this case, at the moment of saturation of the choke L_3, double voltage appears on the active element and the pulsed pump power increases. The reactor L_1 is chosen to be sufficiently large (~ 1 mH) and serves to charge the capacitance C_2.

Figure 4.2 is a photograph of the time dependence of the total pulse current $I_1 + I_2$ flowing through capacitance C_1 and corresponding to the optimal operating mode at $f_{rep} = 15.625$ kHz. The current I_1 passes through its maximum value, after that the current I_2 is activated, the rate of increase of which is higher than that of I_1.

Structurally, the chokes L_2 and L_3 are made in the form of a piece of copper tube with a diameter of $8 \div 10$ mm, on which the ferrite rings are separated, separated by thin fiberglass laminations. The dimensions of the choke are chosen from the saturation condition of the ferrite rings:

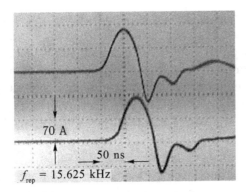

70 A

50 ns

$f_{\text{rep}} = 15.625$ kHz

Fig. 4.2. The pulse shape of the total current $I_1 + I_2$ at $C_1 = C_2 = 1100$ pF.

$$S \cdot n = 0.5 \cdot U_m \cdot t_0 / \Delta B, \qquad (4.1)$$

where n is the number of ferrite rings, S is the cross-sectional area of the ring, U_m is the maximum voltage on the choke, and ΔB is the maximum increment in the magnetic field induction in the choke core.

The values of the chokes L_2 and L_3 are chosen depending on the value of the supply voltage and the type of thyratron used. When using the thyratron TGI-2000/35, when operating on the AE GL-201 for the L_2 choke, rings of the type 2000 HM-20 × 10 × 6 were used. The number of rings was 20 ($S = 5.5 \div 6$ cm²). For thyratron ТГИ-1000/25 the dimensions of this choke were 1.5 times less. For the choke L_3, rings of the type 2000 HM-16 × 10 × 4.5 were used. The number of rings was 168 when working with the thyratron TGI-2000/35 for the active element GL-201 ($S = 22 \div 24$ cm²) and 52 for the thyratonone TGI-1000/25 and the active element Kulon-M (7.2 cm²). Photographs of the design of the L_2 and L_3 chokes are presented in Appendix 3 (Figu. A.3.14), and in Fig. A.3.15 shows photographs of the design of the entire modulator unit for a laser with an AE type GL-201 or Kristall LT-30Cu, operating at a repetition frequency of 15.625 kHz.

When creating a prototype of a projection system with a copper vapour laser, high demands are placed on the development of a high-voltage power supply for this laser with an average output power of more than 4 kW capable of heating active elements such as GL-201 or Kristall LT-30Cu. Appendix 4 gives an example of the development of such a source, in which the following tasks were solved.

1. The power source is a transistor high-frequency voltage converter with an output voltage of up to 7 kV and an average output power of up to 4.5 kW. It is designed in accordance with the full-bridge design and operates at a frequency of 50 kHz with phase control. The high frequency of the conversion makes it possible, on the one hand, to significantly reduce the size of the power elements and the overall dimensions of the power supply, on the other hand, to efficiently filter the interference created by the operation of the thyratron modulator and the transistor converter.

2. The operation of the power supply is controlled from the microcontroller built into the control system of the laser installation. With its help, control of the switching on and off of lasers is carried out, emergency modes are monitored (thyratron failure, lack of cooling, phase failure of the input supply voltage), various parameters of the operation of lasers are monitored and set: the active element heating modes (stepwise pump power rise), pumping power, time cooling, modes of operation of thyratrons, the time of accumulation of active elements and thyratrons is fixed.

3. The power supply provides for stabilization of its output parameters: voltage or power supplied to the load. During the heating of the active element, its resistance varies greatly, which leads to undesirable overloads on the thyratron. The possibility of stabilizing the electric power supplied to the active element allows the heater to be heated more correctly, to remove the loads from the excitation system during the transient process of establishing the conductivity of the active element, and to stabilize the thermal regime of the active element when reaching a given output power level. The feedback loop of the stabilizer also provides for the introduction of signals from optical sensors of medium output radiation power.

Stabilization of the thyratron operating modes in the modulator unit is of great importance for stabilizing the output parameters of the radiation and increasing the reliability of the laser operation. For these purposes, the supply of all auxiliary circuits of the thyratron (voltage, hydrogen generator voltage, auxiliary grid voltage) is provided by separate stabilized high-frequency converters. This makes it possible to reduce the instability (jitter) of the front of the output pulse of the pump voltage and the associated instability of the front of the light pulse, and to reduce the instability of the amplitude of the pump pulses to zero.

All these measures allow achieving stable values of the output optical power level of the laser radiation necessary for the formation of television images. The construction and operation of such a power source for a copper vapour laser are described in more detail in Appendices 3 and 4.

In a copper vapour laser with a plane-parallel resonator, the output radiation is a set of light beams having different angular divergence, depending on the number of complete passes in the resonator [54]. This leads to a broadening of the distribution of the intensity of light in a line on the screen along a coordinate that coincides with the direction of the vertical scanning. The radiation divergence close to diffraction can be obtained by selecting beams, for example, by using an unstable telescopic resonator with an magnification factor $M = R_2/R_1 \geq 200$ [168], where R_2, R_1 are the radii of the mirrors of the resonator. The output power of radiation with a good directionality decreases sharply. A high power, while maintaining the diffraction divergence of the laser radiation, can be obtained in the system by a master laser power amplifier, which was realized in Ref. [76]. In view of the more complex construction of such a laser, its use is justified only for high-definition information display systems and projection of the image onto a large screen. For an information display system with a standard of decomposition of 625 lines in a frame, it is advisable to reduce the requirements for the divergence of laser radiation. One can choose the magnification factor M of the unstable telescopic resonator of a single laser in such a way as to maintain a sufficiently large output power with an acceptable divergence of the light emission. Experimentally, it was found that the optimum value of the magnification factor M for the purposes of using a laser in a television projection device operating with a decomposition standard of 625 lines, 50 fields per second, is $M \simeq 30$ value. Figure 4.3 represents the design of the resonator of such a laser with the active element LT-30 Cu (AE).

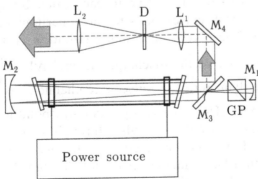

Fig. 4.3. Optical scheme of a laser resonator with an output selector of light beams.

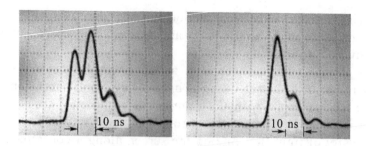

Fig. 4.4. The oscillograms of the pulses of light radiation at the exit from the laser cavity to the diaphragm D (*a*) and after the diaphragm (*b*).

The resonator of the master oscillator is made on a three-mirror scheme of a telescopic unstable resonator with a magnification factor of 30, consisting of spherical concave mirrors M_1, M_2 with a radius- the curvatures $R_1 = 10$ cm and $R_2 = 3$ m and the plane mirror M_3 for outputting radiation with a communication hole of 1 mm diameter. The connection hole of the mirror M_3 is located at the point of common focus of the mirrors M_1 and M_2. The use of a reflecting mirror for radiation output, in contrast to a passing mirror with a reflective meniscus, allows one to get rid of the possible interference of light beams in the output radiation. To obtain linearly polarized radiation, a Glan prism (GP) is placed between the mirrors M_1 and M_3 in the laser cavity.

The output radiation of a laser with this resonator consists of three beams, corresponding to a different number of passes inside an unstable resonator and having a different angular divergence. In Fig. 4.4 *a* is a photograph of the light pulse after the mirror M_3.

The light pulses were recorded with a photovoltaic cathode (FK-19) and displayed on an oscilloscope screen. The radiation pulse consists of spikes shifted relative to each other by approximately 10 ns. The first peak corresponds to the superluminescence of the active element. The divergence of this radiation is determined by the angular aperture of the discharge channel $\Delta\Psi_0 = \dfrac{d_0}{2\cdot l} \approx 10^{-2}$ rad, where d_0 is the diameter of the discharge channel, l is the length of the active element. The other two peaks correspond to one and two complete traversals of radiation pulses through the resonator. Due to the finite time of existence of inversion in the active medium ($30 \div 40$ ns), beams with a large number of passes through the resonator do not have time to form. With the magnification factor $M = 30$, the second light beam corresponding to one cavity pass has a divergence $\Delta\Psi_1 \simeq 3\cdot10^{-4}$ rad, and the third – approximately diffraction divergence. A mandatory attribute of such a laser design is the spatial selector of light beams installed at the output of the resonator, consisting of the lens telescope L_1, L_2 and the diaphragm D. The spatial selector filters the radiation, delaying the light beams with a large angular divergence. Another function of this optical device is to match the diameter of the output beam to the length of the AOM sound line. Figure 4.4 *b* shows a photo of the light pulse after the diaphragm D, which delays the first beam with a large angular divergence. This beam accounts for almost half of the light emission power. With a total light power of 22 W on two lines and a pulse repetition rate of 15.625 kHz, the average power in beams with good divergence is only 12 W at a power consumption of 3 kW. In this case, the generation powers on individual lines are in the ratio $P_{green}/P_{yellow} \simeq 1$.

The total divergence of the laser radiation after the diaphragm was measured by recording the intensity distribution of the light field in the focal plane of the lens with a focal length of 1.6 m. The measured divergence at a wavelength of 510.6 nm at a level of 3 dB was $\Delta\Psi \simeq 1.6 \cdot 10^{-4}$ rad. When using an acousto-optical deflector made of TeO_2 having an aperture of 15 mm and a control frequency band $\Delta f = 50$ MHz, such an angle of divergence of radiation makes it possible to obtain 245 solvable elements in the frame by the Rayleigh criterion. To increase the number of solvable system elements per frame, one needs to increase the scanning angle. This can be done, for example, using an electromagnetic mirror galvanometer.

The operation of the pulse imaging system of the line occurs in such a way that an instantaneous image of the amplitude-modulated ultrasonic field filling the AOM crystal is constructed on one of the coordinates in the plane of the screen, so that high demands are not made on the direction of the light emission from this coordinate. These requirements basically reduce to the possibility of dividing the diffraction orders in the spectral plane. On the other coordinate, in order to obtain a large number of solvable elements, it is necessary to ensure a high directivity of the laser radiation. Such a discrepancy between the requirements for the divergence of a light beam with respect to two coordinates can be used to increase the useful power of the output radiation in a copper vapour laser. For this purpose, in [198], the mirrors M_1 and M_2 of the laser cavity were fabricated in the form of two cylindrical mirrors with the same radii of curvature as for a spherical resonator (see Fig. 4.3). Thus, a resonator with plane-parallel reflecting mirrors was formed on one coordinate corresponding to the line of formation of the line image, and on the other – an unstable resonator with an magnification factor of $M = 30$. The emission of radiation from such a resonator was realized by means of two flat mirrors with a narrow adjustable gap between their faces . The ends of the mirrors forming the slit were mitered at an angle of 45°. To ensure the same phase delay, the mirrors were made of the same thickness and were located on the same guide plane. Comparison of output powers and laser radiation divergences was made in the cases with spherical and cylindrical mirrors and at the same pump level. The measurements showed that, with the laser radiation divergence preserved at one coordinate, the output power of laser radiation when using a resonator with cylindrical mirrors increased by approximately 20% and amounted to 26 W for all light beams. The radiation was selected by means of a cylindrical telescope with a diaphragm in the form of a narrow slit. Thus, the use of a cylindrical, unstable resonator with a large magnification coefficient for only one coordinate in a laser generator on copper vapour allows one to increase the power of the light beam and raise the efficiency of the system.

4.1.1. A copper vapour laser with intracavity acousto-optical control of its spectral and temporal characteristics

The work of adjusting the acousto-optical imaging system using a powerful pulsed laser is associated with difficulties due to the high brightness of the light source. The average output power of the laser generator depends on the mode of heating of the active element. When the heating mode of the copper vapour laser is changed, the relationships between the powers of the spectral components, the duration of the radiation pulses, and the divergences of the light beams [169] change, which prevents the correct tuning of the system. To overcome these difficulties, an acousto-optical control system for the output radiation of a copper vapour laser [170–172] was developed, which allows the amplitude, repetition frequency, and wavelength (λ_1 or λ_2) of laser pulses to be accurately controlled without changing the mode of heating of the active element and direction propagation of laser radiation. This is also important for solving other problems in which a powerful copper vapour laser is used, for example, for precise exposure control in medical

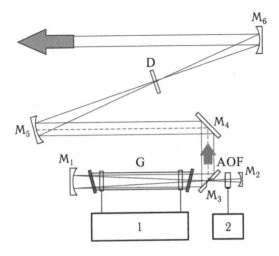

Fig. 4.5. Optical scheme of a laser with intracavity control.

applications (laser surgery, photodynamic therapy, laser endoscopy), laser processing and cutting materials. The control of the light pulses also makes it possible to interrupt radiation when forming a vector graphic image.

Figure 4.5 shows the optical scheme of a copper vapour laser with intracavity control.

The laser resonator is made according to a three-mirror scheme of a telescopic unstable resonator, similar to the one in Fig. 4.3. To control the laser radiation between the mirrors M_2 and M_3 inside the resonator, instead of the Glan prism, an acousto-optical tunable filter (AOF) made of a paratellurite crystal (TeO_2) is installed inside the resonator. The AOF uses the wide-angle non-collinear geometry of the acousto-optical interaction proposed in Ref. [100].

Figure 4.6 shows the geometry of the acousto-optical interaction in a tunable filter with the diffraction of light in the +1st order and the arrangement of the AOF inside the copper vapour resonator. The angle between the wave vector of the incident light k_i and the [001] axis in the TeO_2 crystal is 23.5°, and the angle between the wave vector of the ultrasonic wave q and the [110] direction is 10°. With this geometry, the acousto-optical tunable filter is not critical with respect to the adjustment of the Bragg angle when the wavelength of the incident light changes.

The formation of controlled output radiation is as follows. The intracavity light beam passes through the communication hole of

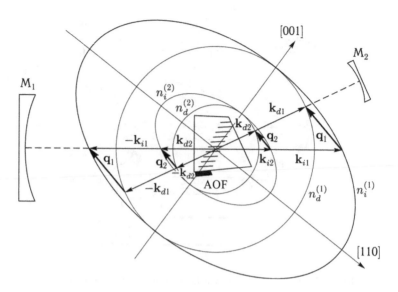

Fig. 4.6. The geometry of the acousto-optical interaction in the AOF and the scheme of its location inside the copper vapour laser cavity. $\mathbf{k}_{i,d(1,2)}$ are the wave vectors of the incident (i) and diffracted (d) radiation at wavelengths of green (1) and yellow (2) colors; $\mathbf{q}_{1,2}$ are the wave vectors of the ultrasonic wave for interaction with the wavelengths of green (1) and yellow (2) colours.

the mirror M_3 and falls on the AOF. In the AOF, the power of a piezoelectric transducer from $LiNbO_3$ is excited by an ultrasonic wave with frequencies Ω_1 or Ω_2 corresponding to the Bragg diffraction conditions for light with wavelengths $\lambda_1 = 510.6$ nm or $\lambda_2 = 578.2$ nm. For each of these light waves, the light deviates into +1st and −1st orders of diffraction. The mirror M_2 is tuned so that the light beams reflected from it and the second time diffused on the sound are returned back to the resonator of the laser, reflected from the mirror M_1, amplified as a result of double passage through the active medium, and with the aid of M_3, they are extracted outward. In AOF from TeO_2, anisotropic diffraction of light by sound is used, and normal types of light waves are waves with linear polarization. Figure 4.6 shows the diffraction in +1st order. With this diffraction, the light emission at the output from the laser is linearly polarized in the plane of the figure. As a result of adjusting the mirror M_2 to the +1st or −1st order of diffraction, the linear polarization of the radiation at the output from the laser can change to an orthogonal one, and the radiation frequency after each cavity pass is shifted by $2\Omega_1$ or $2\Omega_2$. Before the laser pulse is generated, the radio pulses with carrier frequencies Ω_1 or Ω_2 are separately or simultaneously fed to the AOF piezoelectric transducer. In this case, feedback is turned on in the laser resonator and light pulses are formed at its output with a given length wave. The frequencies Ω_1 and Ω_2, corresponding to λ_1 and λ_2, are 157 MHz and 133 MHz. The control unit controls the time of appearance of each light pulse, and the amplitude of the light pulses can be changed by changing the amplitude of the radio pulses at the input of the AOF.

To suppress the superluminescence in the absence of control pulses, as well as for angular selection, the output radiation is passed through a spatial filter in the form of a diaphragm D, installed in the focal plane of an additional mirror telescope formed by mirrors M_5 and M_6. The AOF from TeO_2 serves as an effective polarizer of laser radiation. Comparison of the output powers of the laser with AOF from TeO_2 and the Glan prism installed at the same place in the resonator showed that for one and the same pump power, the average output power of a copper vapour laser remains practically the same with an AOF diffraction efficiency of about 80%.

Figure 4.7 shows an acousto-optical tunable filter on the adjustment fixture, with which it was fixed inside the laser cavity.

Thus, the use of a system of intracavity control of the output radiation of a copper vapour laser based on AOF makes it possible to control the wavelength and amplitude of each laser light pulse without changing the mode of heating of the active element and the direction of light propagation. This greatly simplifies the work on the adjustment of the optical system of the television projector, and also expands the possibilities of other applications of the copper

Fig. 4.7. Acousto-optical tunable filter for a copper vapour laser on the adjustment device. 1 – AOF, 2 – matching device.

(a)

Fig. 4.8 (*a, b*). The operation of the laser system with intracavity switching wavelength generation.

vapour laser, in particular, for the technological application of both the main radiation of a copper vapour laser and a converted one, for example, in the UV region of the spectrum.

As an example of the operation of a copper vapour laser with intracavity control of the spectral characteristics of the radiation Fig. 4.8 shows the operation of the laser installation when forming the logo of a well-known company on the wall of the building of the former City Council in St. Petersburg.

4.1.2. Control of the parameters of the light beam in the the oscillator–amplifier of laser radiation system

The active medium of metal vapour lasers has a high gain, which makes it possible to use these lasers as amplifiers for the brightness of light beams. This property of the active medium opens up great possibilities for obtaining powerful laser radiation with high temporal and spatial characteristics, which are necessary for an acousto-optical projection system of imaging on a large screen. At the same time, the power losses that normally occur when a light beam with the required parameters are formed can be compensated for by amplification in a laser amplifier. In the master oscillator, it is possible to use laser tubes of small dimensions for which a small-sized pumping system can be created. The small length of the cavity of the master laser, as well as the small inductance of the discharge circuit, make it possible to form light pulses with a duration much shorter than when using active elements of large dimensions, and thereby increase the resolution and contrast of the image being formed.

In work [32], studies were made of the possibility of obtaining powerful light beams with the characteristics required for a pulsed system of image formation. A laboratory installation of a generator and amplifier of laser radiation was created. The installation was a small-sized master laser on copper vapour with the active element Kulon-SM (length of the AE $l = 16$ cm, channel diameter $D = 8$ mm) and power amplifier of light beams based on the laser active element GL-201 ($l = 95$ cm, $D = 20$ cm). The resonator of the laser oscillator corresponded to the optical scheme in Fig. 4.3. It used an unstable telescopic resonator with a magnification factor $M = 50$ and a radius of curvature of the mirrors M_1 and M_2, respectively, of 2 cm and 1 m. The radiation from the resonator was outputted by means of a flat mirror with a communication hole 0.5 mm in diameter. To obtain linearly polarized radiation, the Glan prism was placed in the laser cavity. The switching of electric energy was carried out by the TGI 3-500/16 thyratron according to the scheme shown in Fig. 4.1 *a*. The operation of this laser was performed with repetition frequencies of 15.625 kHz and 28.125 kHz corresponding to the frequency of horizontal sync pulses for a conventional television standard and an enhanced definition standard with the number of lines in frame 1125. Depending on the frequency of repetition of the pump pulses, the values of the working and aggravating capacities were chosen equal to: $C_0 = 1100$ pF, $C_1 = 220$ pF at $f_w = 15.625$ kHz and $C_0 = 800$ pF,

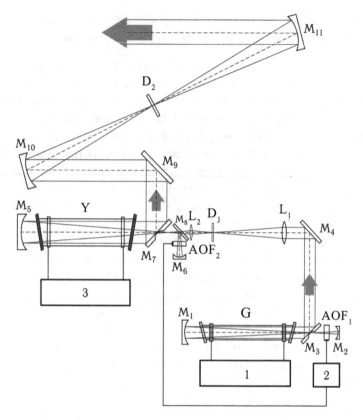

Fig. 4.9. Optical scheme of a laser system of two coupled copper vapour lasers.

$C_1 = 160$ pF at $f_w = 28.125$ kHz. With an average power consumption of 500 W from the rectifier, the average output power of linearly polarized laser radiation on the green and yellow lines was 0.6 W at 15.625 kHz and 1.1 W at a frequency of 28.125 kHz, and the pulse duration after the superluminescence filtration was equal to 5 ns by the level of half power. The divergence of the light beam arriving at the input of the laser amplifier amounted to $0.8 \cdot 10^{-4}$ rad.

The laser radiation amplifier operated at a repetition rate of 15.625 kHz. The launch of the amplifier was carried out synchronously with the launch of the laser generator through a variable delay block. The region of relative delay between the two triggering pulses, within which the light pulse was amplified, was 25 ns. The delay between the generation pulses on the green and yellow lines of the laser with an unstable resonator was practically absent, and for the amplifier it was about 20 ns, so choosing the delay allowed an effective tuning of the radiation wavelength at the output of the laser amplifier.

The average output power of the laser radiation after the amplifier was 7 W (at P_{in} = 0.4 W) at a wavelength of 510.6 nm, which is half the output power of the active element used in its operation with a stable resonator in the generator mode at the same pump parameters. Thus, the power parameters of the master oscillator in this setup did not allow to remove the entire inversion from the active element of the amplifier in one pass of the light pulse.

4.1.3. A system of two lasers on copper vapour with the injection of a light beam into a powerful laser

In order to increase the output power of laser radiation with a good directivity, a study was made of a system of lasers with coupled resonators, similar to that which was considered in Ref. [173]. Figure 4.9 shows the optical scheme of the laser system.

The laser system consisted of a relatively low-power copper vapour laser (G) with the active element Kulon (LT-3Cu) and an average power level laser (Y) with the active element LT-30Cu. The resonators of both lasers were assembled according to the laser scheme with an unstable resonator and intracavity control, considered in section 4.1.1.

In the first laser, an unstable telescopic resonator with $M = 30$ and a curvature radius of the mirrors R_1 = 1.5 m and R_2 = 5 cm was used. In the second, more powerful laser, $M = 15$, R_5 = 3 m, R_6 = 20 cm. In the resonator of a powerful laser Y between its output mirror M_7 and a mirror M_6 at an angle of 45°, a beam splitter M_8 was placed which deflected or transmitted 50% of the incident light at wavelengths 510.6 nm and 578.2 nm. The input of radiation from a low-power laser G into the resonator of the Y laser. Using the telescopic systems from the lenses A_1 and A_2, the output radiation was coordinated with the light mode of the Y laser. The diaphragm D_1 carried out the selection of the light beams after the laser G, and prevented the radiation from the powerful laser from being brought back into the resonator of the first laser. Both lasers worked synchronously at a frequency of 15.625 kHz. The delay between the triggering pulses of the first and second lasers could be smoothly controlled within 50 ns, which covered the inversion time in the high-power laser, and was maintained with an accuracy of ±1 ns. The AOF in lasers worked synchronously and simultaneously transmitted

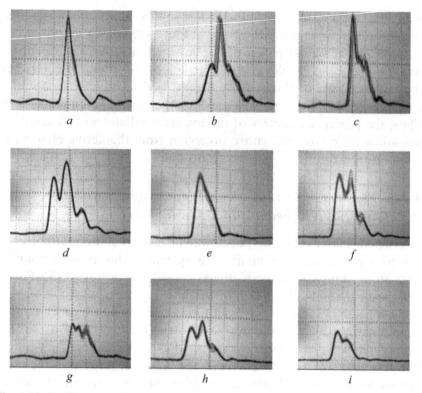

Fig. 4.10. Oscillograms of light pulses at the output of an oscillator–amplifier system with coupled resonators. In these photographs, one horizontal cell corresponds to 10 ns.

two wavelengths of radiation. The average generation power of the laser G after the diaphragm D_1 was 2.5 W, and the power passed through the resonator of the Y laser was 1 W.

Figure 4.10 shows photographs of the oscillograms of the light pulses at the output of a powerful Y laser.

Figure 4.10 *a* shows the light pulse from the laser G when the active element of the laser Y is not yet heated. The divergence of this radiation is approximately equal to $2 \cdot 10^{-4}$ rad. As heating continues, superluminescence (*b*) appears in the output radiation before the pulse from the master laser, which, when the delay of the input pulse is adjusted, can go into stimulated emission (*c*) with a divergence of the input signal. Figure 4.10 *b* and *c* shows the 'jitter' between the pulses of radiation and current through the active element Y (see Fig. 4.2) at the initial stage of its heating, approximately equal to 2 ns.

Figure 4.10 *d* shows the radiation pulse only from the powerful laser to the diaphragm D_2 with the overlapped emission of the first laser. As already mentioned above, the radiation pulse has a spike character associated with a different number of radiation passes through the resonator. The first peak corresponds to the emission of superluminescence and is completely filtered by the diaphragm D_2, with only 13 W remaining of the 26 W total output power after the diaphragm.

When injection of the input laser beam is resumed and when the luminescence peak is temporarily matched with the peak of the input pulse the entire power of the superluminescence radiation of a high-power laser goes into the radiation stimulated by the first laser with a good directivity, which passes through the diaphragm D_2. Figures 4.10 *e* and *f* show the pulses of light transmitted through this diaphragm. Figure 4.10 *e* corresponds to the case of overlap of the reflecting mirror M_6. In this case, the Y laser acts as a two-pass power amplifier of the input beam. There is no spike structure in the radiation pulse, which, apparently, is due to 'mixing' and competition of amplification of the input beams in the amplifier. In the steady-state regime, the average output power of laser radiation in this case was 28 W at a ratio of 1:1 between green and yellow components of radiation. With a closed resonator of a powerful laser (mirror M_6 open), a spike structure appears in the output pulse of the radiation (see Fig. 4.10 *f*), which is associated with the intrinsic total radiation passages in the resonator of this laser, also having a good directivity. The duration of the light pulse in this case increases slightly, and the average radiation power increases to 31 watts. Thus, the inversion of the active medium turns almost completely into radiation with a divergence acceptable for the information display system. Figures 4.10 *g, h, i* reflect the shape of laser radiation pulses at the maximum output power of the system on two spectral lines: *g* is the radiation pulse at the input Y, *h* is the output pulse of radiation with an open resonator of a powerful laser, and *i* is the output pulse of radiation with a closed resonance a powerful laser.

A great advantage of the considered system of two lasers, one of which is of low power, is the possibility of a significant increase in the output power of laser radiation with a good directionality at relatively small additional costs, in comparison with a laser generator on one active element. Thanks to the acousto-optical filters installed in the resonators of each laser, the laser system shown in Fig. 4.9, allows to control the time and spectral characteristics of its radiation without changing the modes of heating of the active element. Pulses on green and yellow wavelengths are fed from different laser transitions and do not compete with each other. With the AOF switched off, only superluminescence is present in the output radiation, which is retained by the diaphragms D_1 and D_2. The AOF should work synchronously. To increase the suppression of stimulated emission when the output radiation is turned off, the laser pulse control system delays the pulse of the amplifier's startup by a certain amount (5÷10 µs) relative to the oscillator starting pulses, so that the superluminescence pulses of the oscillator enter the amplifier at a low level of inversion of the active medium.

4.2. Production of a blue line in a titanium–sapphire laser with pumping from a copper vapour laser and transformation of radiation into a second harmonic

To create a full-colour information display system with a pulsed way of forming a line, it is necessary to have 3 pulsed laser light sources with wavelengths located in the green, red and blue regions of the spectrum. A gold vapour laser with a wavelength of 628.3 nm is successfully used for gas lasers on self-terminating metal vapour transitions. The average generation power from one active element of the Kristall type reaches 6÷10 W. In addition, powerful pulsed radiation with a 'red' wavelength can be obtained in a dye laser, which is pumped by a copper vapour laser. As for the blue region of the spectrum for this type of laser, effective and powerful generation could not be realized in it, despite the intensive work carried out in the world during the last 30 years. In this connection, it is of interest to investigate the feasibility of using a copper vapour laser radiation to produce pulsed radiation at a wavelength of blue.

It is theoretically known that this transformation can be achieved most effectively in two ways: first, as a result of parametric transformation in nonlinear media, followed by doubling the frequency of the signal or idle wave, and secondly, as a result of doubling the frequency of the laser radiation on titanium sapphire (Al_2O_3:Ti^{3+}).

The absorption curve of the active medium of a titanium–sapphire laser has a maximum in the region of 500 nm [174], therefore, in most studies, lasers using argon lasers (515, 488 nm) and a second harmonic of the Nd:YAG laser (532 nm) are used to pump it. A few works on the coherent pumping of an Al_2O_3:Ti^{3+} laser with a copper vapour laser [175] have not been further developed. The level of output power and efficiency achieved in them is noticeably inferior

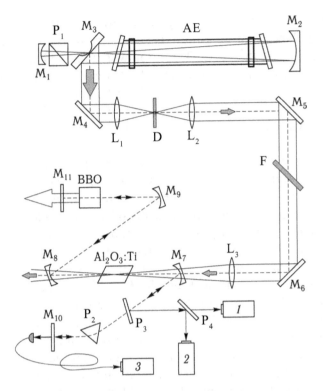

Fig. 4.11. The scheme of the experimental setup for studying the efficiency of pumping a titanium–sapphire laser by a copper vapour laser and further converting radiation into the blue region of the spectrum.

to the similar characteristics of the systems mentioned above. In addition, negative factors that may interfere with the effective use of copper vapour lasers include a large output beam diameter that makes it difficult to reconcile the pump field and the mode of the Al_2O_3:Ti^{3+} laser resonator, and the presence of two spaced lines in the emission spectrum ($\lambda_1 = 510.6$ nm and $\lambda_2 = 578.2$ nm) with strongly differing values of the absorption cross section ($\sigma_1 = 8.3 \cdot 10^{-20}$ cm^2, $\sigma_2 = 3.8 \cdot 10^{-20}$ cm^2).

The aim of the studies carried out in Ref. [176, 177] was to compare the pumping efficiency of the Al_2O_3:Ti^{3+} laser with two different spectral lines of the copper vapour laser radiation, and also develop a promising scheme for converting the radiation into the blue region of the spectrum. Figure 4.11 shows the scheme of the experimental setup. As the active element of the AE, a separate gas-discharge tube of the GL-201 type was used. The resonator of a copper vapour laser was assembled according to the scheme of an unstable telescopic resonator (Fig. 4.3) with a coefficient $M = 37$. The linear polarization of radiation necessary to reduce the losses during pumping of the $Al_2O_3:Ti^{3+}$ laser was achieved by installing a Glan polarization prism (P_1). The diameter of the light beam at the output of the telescopic system of the L_1 and L_2 lenses was 15 mm. Diaphragm D carried out a spatial selection of the radiation of a copper vapour laser. The measured radiation divergence was approximately $3.5 \cdot 10^{-4}$ rad.

The copper vapour laser operated at a frequency of 15.625 kHz and generated light pulses of 20 ns duration. The average power of the pump radiation, measured after the M_7 mirror, reached 7 W at a ratio of the green and yellow components of 3/4. The interchangeable dichroic mirrors Φ made it possible to change the relations between the green and yellow components of the pump spectrum, which was required to determine their relative effectiveness.

Using a lens L_3 with a focal length of 20 cm, the pump radiation passing through the mirror M_7 of the titanium–sapphire laser resonator was focused inside the $Al_2O_3:Ti^{3+}$ crystal. Measurement of the distribution of the intensity of the light emission in the waist of the focusing lens L_3 has shown that the light spot has the shape of a circle whose diameter is equal to 100 μm for the green and yellow components, which approximately coincided with the dimensions of the mode of the titanium–sapphire laser resonator.

The titanium–sapphire laser resonator was made according to a standard Z-shaped scheme with compensation for astigmatism. An active element of $Al_2O_3:Ti^{3+}$ with a diameter of 5 and a length of 15 mm with Brewster ends was established between spherical mirrors M_7 and M_8 with radii of curvature of 10 cm. The transmission of the active element was $T_{P_1} = 1.8\%$ at a wavelength $\lambda_1 = 510.6$ nm and $T_{P_2} = 16.8\%$ at the wavelength $\lambda_2 = 578.2$ nm. To compensate for the astigmatism of the resonator, the mirrors M_7 and M_8 were installed at an angle of 15° to the incident radiation. The distance between these mirrors was chosen on the basis of the condition of falling into the region of stable generation with respect to two coordinates [178] and equaled 12.5 cm. When investigating the pumping efficiency, flat mirrors M_9 and M_{10} were installed in the resonator. The distances between the mirrors M_8 and M_9, and also M_7 and M_{10} were equal to 31 cm. The reflection coefficients of all the mirrors in the center of the tuning range exceeded 99%. During the experiments, two sets of mirrors were used for the ranges 680÷850 and 850÷980 nm. The

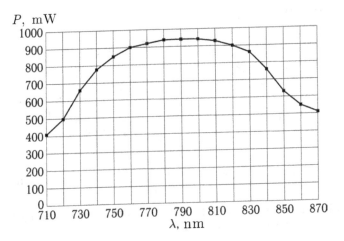

Fig. 4.12. The dependence of the average power of the Ti:Al$_2$O$_3$ laser on the wavelength at 16% load of the resonator for the first set of mirrors and pumping by two lines of a copper vapour laser.

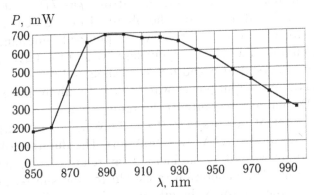

Fig. 4.13. Dependences of the average power of the Ti:Al$_2$O$_3$ laser on the wavelength at 16% of the load of the resonator for the second set of mirrors and pumping by two lines of a copper vapour laser.

tuning of the generation wavelength was carried out by the Brewster glass prism P$_2$, and the radiation output from the resonator was by means of a thin plane-parallel glass plate P$_3$ installed at an angle to the axis of the resonator of the titanium–sapphire laser. The average generation and pumping powers were measured using a power meter *1* (IMO-2). The time characteristics of the radiation were recorded by a photoelectric cathode *2* (PK-19). The radiation along the optical fiber was directed to the input of the spectrometer *3* (SFK-601) for measuring the wavelength.

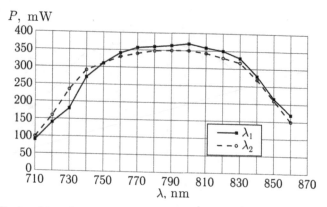

Fig. 4.14. Tuning dependences of a Ti:Al$_2$O$_3$ laser with separate pumping by radiation with λ_1 = 510.6 nm and λ_2 = 578.2 nm.

The tuning characteristics were studied at a 16% load of the resonator, given by the rotation of the plate P$_3$. The generation area realized with the use of two sets of mirrors was 710÷970 nm.

Figures 4.12 and 4.13 show the obtained tuning dependences of Al$_2$O$_3$:Ti^{3+} for two sets of mirrors when pumped simultaneously by two wavelengths of a copper vapour laser with a total average power incident on a crystal of 6 and 6.8 W, respectively.

The maximum output power was achieved with an increase in the load to 40% and at the centre of the tuning characteristic, the value for the first set of mirrors was P = 1.32 W (λ = 790 nm), and for the second set of P = 1.2 W (λ = 900 nm). The duration of the laser pulse at 900 nm was 40 ns and had a delay relative to the pump pulse of about 40 ns. The generation threshold for the short-wave region was 1 W, for the long-wavelength region it was 2 W.

The differential conversion efficiency at the centre of the gain line of a Ti:Al$_2$O$_3$ laser at a wavelength of 770 nm was 26.4%, and at a wavelength of 900 nm it was 24%.

To compare the pumping efficiencies of different laser wavelengths on copper vapors, the tuning dependences of the Ti:Al$_2$O$_3$ laser were measured with separate pumping by radiation with λ_1 = 510.6 nm and λ_2 = 578.2 nm for the first set of mirrors. These dependences are shown in Fig. 4.14.

The pumping power on each of the lines was 3 W. As can be seen from the graphs given, despite the strong difference in the absorption cross sections, both the pump conversion efficiency and the tuning region in both cases are practically identical.

The results obtained can be explained by using the expression for the differential efficiency [179, 180]:

$$\eta = \frac{\lambda_p}{\lambda_g} \cdot \frac{T}{(T+\delta)} \cdot (1-T_p), \qquad (4.2)$$

where λ_p, λ_g are the wavelengths of the pump and generation, T is the transmission loss through the laser mirrors, δ is the passive loss in the resonator, and T_p is the transmission of the crystal at the pump wavelength. From (4.2) it is easy to obtain a ratio for the relative pumping efficiency at wavelengths λ_{P_1} and λ_{P_2}:

$$\frac{\eta_1}{\eta_2} = \frac{\lambda_{p_1}}{\lambda_{p_2}} \cdot \frac{(1-T_{p_1})}{(1-T_{p_2})}. \qquad (4.3)$$

Substituting in (4.3) the values of the quantities corresponding to the experimental conditions, we obtain $\frac{\eta_1}{\eta_2} = 1.04$, which agrees well with the results of the experiment. Physically, this means that in the present case of two-frequency pumping, mutual effects are compensated for due to the difference in the Stokes shift and absorption in the active medium.

To obtain a second harmonic from a Ti:Al$_2$O$_3$ laser operating at 900 nm, one of the arms of the Z-shaped resonator changed: the plate P$_3$ was removed and the plane reflecting mirror of M$_9$ was replaced by a semiconfocal system of mirrors M$_9$, M$_{11}$ reflecting the radiation at a wavelength of 900 nm. A crystal of BBO with enlarged ends for the main and second harmonics (λ = 450 nm) was established

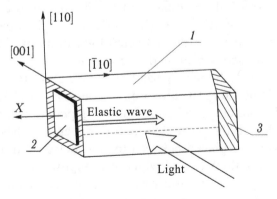

Fig. 4.15. AOM design with direct excitation of a slow shear wave.

between the mirrors M_9 and M_{11} near the plane mirror. The second harmonic of the fundamental radiation was output through a plane mirror M_{11}.

The power was 400 mW at a wavelength of 450 nm with a generation duration of 40 ns and a pulse repetition rate of 15.625 kHz, which makes it possible to expect a further higher average radiation power in the blue region when using a much more powerful vapour laser copper.

4.3. The design of acousto–optical modulators and deflectors for a pulse information display system

As already noted in Ch. 2, the most suitable material for the acousto–optical modulator of an imaging system with a pulsed laser is at present a paratellurite TeO_2 crystal. This material can also be used with success in the manufacture of an acousto-optical deflector for deflecting a light beam along a frame.

For the first time, the construction of an acousto-optical modulator from TeO_2 was proposed in Ref. [129] and has since been the most common.

In this construction (Fig. 4.15), a slow transverse acoustic wave is excited along the [110] direction of the crystal by a transverse-wave piezotransducer from $LiNbO_3$, for example, an X-cut located on the surface (110) of the sound line. Acoustic impedances of the materials of the sound transmission and piezoelectric transducer $Z = \rho \cdot v$, where ρ is the density, v is the speed of sound, are very different under such excitation, so their ratio for the piezoelectric transducer of the X-cut from $LiNbO_3$ is 0.165. This results in a narrow bandwidth of the piezoelectric transducer and thus worsens such parameters of the acousto-optical device as the number of allowed angular positions of the deflected light beam and the product of the intensity of the deflected light beam by the frequency bandwidth of the acousto-optical interaction.

To provide a wide frequency band of excited slow transverse waves propagating along the (110) direction, the authors of [181] proposed to perform the matching of the acoustic impedances of the sound guide materials and the piezoelectric transducer. For these purposes, an indium layer was placed between the piezoelectric transducer and the sound line, the thickness of which was approximately one quarter of the length of the ultrasonic wave in In, chosen from the central frequency of the acousto-optical interaction band.

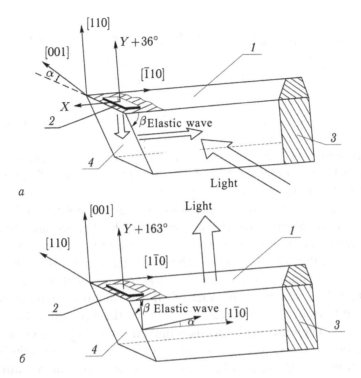

Fig. 4.16. The construction of acousto-optical devices with the transformation of the types of elastic waves on the lateral side of the sound line

The drawbacks of this modulator design are, firstly, the complexity of the technology of its manufacture, consisting in the necessity to withstand the required thickness of the intermediate intermediate layer from In and the high accuracy of the orientation of the face of the sound transmission line on which the piezoelectric transducer is placed. Secondly, when the temperature in the material of the sound line changes, optical inhomogeneities arise due to the large difference in the coefficients of linear expansion in the rigidly connected planes (110) of the TeO_2 and (110) piezoelectric transducer from $LiNbO_3$. Thirdly, in such a design it is impossible to correct the direction of propagation of an ultrasonic wave in the modulator's sound line after it is manufactured. Thus, the error in the orientation of the face on which the piezoelectric transducer is placed is turned by a small angle ϕ relative to the **[110]** axis in the **(001)** plane of the TeO_2 crystal, causing the energy flow of the elastic wave to deviate in the finished device by an angle $\tilde{\omega}$ in the plane indicated, which is the plane of the optical aperture of the device, and

$$\text{tg}\tilde{\omega} \simeq \left(\frac{1+\phi}{1-\phi}\right)\cdot\frac{(c_{12}+c_{66})\cdot(c_{11}+c_{12})+4(c_{11}+c_{12})\cdot\left(c_{66}-\frac{c_{11}-c_{12}}{2}\right)\cdot\phi}{(c_{12}+c_{66})\cdot(c_{11}-c_{12})-4(c_{11}+c_{12})\cdot\left(c_{66}-\frac{c_{11}-c_{12}}{2}\right)\cdot\phi}, \tag{4.4}$$

where c_{ij} are the elastic constants of TeO_2. It follows from (4.4) that for $\phi = 0.5°$ the angle $\tilde{\omega} \simeq 25°$.

This leads to the fact that at high D/h ratios, where h is the height of the piezoelectric transducer, and D is the size of the optical aperture in the ultrasonic propagation direction, the acousto-optical interaction will be realized only with part of the light flux incident on the AOM. Similar results are obtained by the wedge shape of the intermediate bonding layer from In. In addition, the appearance of optical inhomogeneities in the acoustic conductor when the temperature is changed due to the appearance of mechanical stresses leads to a decrease in the number of allowed angular positions of the deflected light beam in comparison with the theoretically possible, as well as to a decrease in the acousto-optical interaction efficiency due to a change in the specified orientation polarization of the light beam.

To eliminate these shortcomings of acousto-optical modulators and deflectors made from TeO_2, the design of these devices was developed in Refs. [124, 182] for use in the information display system with a pulsed laser. Figure 4.16 *a, b* presents two versions of this design.

In these devices, the problem of matching acoustic impedances is solved by converting the types of vibrations of a sound wave with large values of velocities into modes with small values at the interface between two media [183]. Paratellurite has a large difference in the extreme values of shear and longitudinal wave velocities. By exciting elastic waves with high velocities in the acoustic waveguide, we thereby approximate the acoustic impedance of the sound line to the acoustic impedance of the piezoelectric transducer and increase the broadband of the device.

Of great importance in the manufacture of acousto-optical devices is the choice of the cutoff of a single crystal used for a piezoelectric transducer, and also the mutual orientation of the contacting planes of a sound line and a piezoelectric transducer. The choice of these components should be made with allowance for obtaining the smallest difference in the temperature coefficients of their linear expansion, since otherwise, as a result of thermal compression welding, residual stresses appear that have the character of tensile forces that act from the side of the piezoelectric transducer to the sound line and lead to the appearance of optical inhomogeneities in the sound pipe. A comparative analysis of the characteristic surfaces of the temperature coefficients of linear expansion for various sections of $LiNbO_3$ used to efficiently excite acoustic oscillations in the [110] TeO_2 direction showed that the smallest difference in the coefficients of linear expansion is observed for the (110) TeO_2 plane and the $Y_{+36°}$ $LiNbO_3$, which effectively excites a longitudinal ultrasonic wave, with such a mutual orientation, when the X axis of the piezoelectric transducer is oriented along the $[1\overline{1}0]$ TeO_2 direction. Thus, in the case of the device shown in Fig. 4.16, *a*, it is possible not only to reconcile the acoustic impedances of the piezoelectric transducer and the sound line, i.e., to abandon the additional quarter-wave matching layers, but also to significantly reduce the residual stresses in the sound conductor due to the use of a more linear expansion of the piezo- converter.

Operation of the device shown in Fig. 4.16, *a*, proceeds as follows. The longitudinal ultrasonic wave is excited by the piezo-converter (*2*) of the $Y_{+36°}$-cut of $LiNbO_3$ in the sound line (*1*) and propagates along the $[1\overline{1}0]$ axis. The ratio of the acoustic impedances of the sound transmission material and the piezoelectric transducer in this case is 0.794, while for a slow shear wave in this direction and the piezoelectric transducer of the X-cut, this ratio is 0.165. A longitudinal ultrasonic wave falls on the ultrasound-reflecting facet of the sound pipe (*3*), which makes up with the face on which the piezoconverter is placed, the dihedral angle β calculated by the formula:

$$\mathrm{tg}\beta = v_L \left[\left(\frac{c_{11}+c_{12}}{2} \cdot \cos^2 \alpha + c_{44} \cdot \sin^2 \alpha \right) \Big/ \rho \right]^{-\frac{1}{2}}, \qquad (4.5)$$

where c_{ij} are the elastic constants of TeO_2, ρ is the density of TeO_2, v_L is the velocity of longitudinal elastic waves in TeO_2 in the [110] direction, α is the angle between the [001] crystallographic axis in TeO_2 and the edge of the dihedral angle β.

Two elastic waves appear in reflection from this face: a slow quasi-transverse and a quasi-longitudinal wave, and the conversion coefficient of the incident longitudinal wave into the slow quasi-transverse wave is 0.9; i.e., almost all of the energy of the longitudinal ultrasonic wave is transformed into the energy of the quasi-transverse wave. The angle α is chosen to be zero when the device is operating as an AOM and $\alpha \cong 5 \div 6°$ when operating as an AOD. In the second case, the frequency at which the two-phonon interaction occurs is outside the band of working frequencies of the deflector and there is no dip in the amplitude–frequency characteristic of the device. An elliptically polarized light wave is incident on the Bragg angle sound line to the [**001**] axis of the sound line in the (**110**) plane, which is the scattering plane. The polarization of the light wave is established in such a way that the regime of broadband anisotropic diffraction of light by sound is realized. The energy of ultrasonic waves interacting with light is absorbed by the absorber (*3*).

The device shown in Fig. 4.16 *b* differs from the device of the first type in that a shear ultrasonic wave propagating along the [001] axis of the sound line is excited in the TeO_2 sound line, the displacement vector of which is directed along the [110] axis. The elastic wave transducer in this case is a plate of the $Y_{+163°}$-cut $LiNbO_3$. The ratio of the acoustic impedances of the materials of the sound line and the piezoelectric transducer in this case is 0.608. The dihedral angle β for a given type of modulator is calculated from the formula (4.5), in which instead of v_L it is necessary to substitute the value of the velocity of transverse elastic waves along the [001] direction and with the polarization along the [110] axis in TeO_2.

The acousto-optical modulators and deflectors shown in Fig. 4.16 have, in addition to what has been said above, an advantage over conventional devices that the change in the direction of propagation of the energy flow of the elastic wave in these devices with respect to the direction of propagation of the slow ultrasonic wave resulting from the inaccuracy of the orientation of the vertex face on which the piezoelectric transducer is placed, or tapering of the tie layer, can be corrected at the final stage of manufacturing of these devices after welding and scraping the transducer. Indeed, a wave of the required type and direction of propagation is formed as a result of reflection from the side face of the sound line. In this case, the wave vectors of the incident $\mathbf{q}_i$ and reflected $\mathbf{q}_d$ sound waves, as well as the normal to the reflecting face $\mathbf{n}$, lie in one plane, and the velocities of the incident v_i and reflected v_d waves satisfy the relation

$$v_i \cdot \sin \varphi_d = v_d \cdot \sin \varphi_i, \qquad (4.6)$$

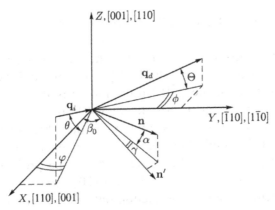

Fig. 4.17. A coordinate system for determining the angles Θ and $\Box$ of a reflected elastic wave

where φ_i and φ_d are the angles of incidence and reflection of sound waves, respectively, with the orientation of the reflecting plane unchanged. The change in the direction of propagation of the incident elastic wave causes the angles of incidence and reflection, as well as the values of the velocities of the incident and reflected waves, to change. However, the change in the direction of propagation of the reflected wave can be compensated if, at the same time, the orientation of the reflecting face of the sound line is changed to the corresponding angle, so that condition (4.6) is satisfied for the new direction and velocity of the incident wave. Such a correction can be made at the final stage of the device manufacturing by changing the orientation of the reflecting face of the sound line by grinding it while observing the pattern of the ultrasonic field in the transmitted light.

To estimate the value of the necessary correction, consider the coordinate system X, Y, Z (Fig. 4.17), which coincides with the crystallographic axes [110], [110], [001] TeO_2 for the modulator shown in Fig. 4.16 a, and with the crystallographic axes [001], [1$\bar{1}$0], [110] for the modulator in Fig. 4.16 b.

For the device shown in Fig. 4.16 a, the relations between the angles characterizing the propagation directions of the incident and reflected elastic waves, as well as the orientation of the normal to the reflecting face, are:

$$\cos\Theta\cdot\cos\theta\cdot\sin\alpha\cdot\cos(\varphi-\phi)=\sin\Theta\cdot\cos\alpha\cdot\cos\theta\cdot\sin(\beta_0+\gamma-\varphi)+$$
$$+\cos\Theta\cdot\cos\alpha\cdot\sin\theta\cdot\cos(\beta_0+\gamma-\phi) \tag{4.7}$$

$$\cos\varphi_i=\cos\theta\cdot\cos\alpha\cdot\cos(\beta_0+\gamma-\varphi)+\sin\alpha\cdot\sin\theta, \tag{4.8}$$

$$\cos \varphi_d = \cos \Theta \cdot \cos \alpha \cdot \sin(\beta_0 + \gamma - \phi) + \sin \alpha \cdot \sin \Theta, \qquad (4.9)$$

where β_0 is the angle of incidence for the case when the incident elastic wave propagates along the [110] direction, and the reflected wave propagates along the [$\bar{1}$10] direction; θ is the angle between the vector $\mathbf{q}_i$ and its projection onto the (001) plane; φ is the angle between the projection of the vector $\mathbf{q}_i$ on the (001) plane and the [110] axis; Θ is the angle between the vector $\mathbf{q}_d$ and its projection onto the (001) plane; ϕ is the angle between the projection of the vector $\mathbf{q}_d$ on the (001) plane and the [$\bar{1}$10] axis; α is the angle between the normal n and its projection onto the (001) plane; γ is the angle between the projection of the normal n onto the (001) plane and the direction $\mathbf{n}'$, which is the angle β_0 with the [110] axis in the (001) plane of the TeO$_2$ crystal.

Approximate expressions for the phase velocities of the incident quasilongitudinal and reflected quasi-shear elastic waves at small angles φ, θ, Θ, ϕ for the TeO$_2$ crystal have the form:

$$
v_i \simeq v_{i_0} \Bigg\{ 1 - \frac{(c_{11}+c_{12})\left(c_{66} - \dfrac{c_{11}-c_{12}}{2}\right)}{(c_{12}+c_{66})\left(c_{66} + \dfrac{c_{11}+c_{12}}{2}\right)} \cdot \varphi^2 +
$$

$$
+ \frac{c_{44}-c_{66} - \dfrac{c_{11}+c_{12}}{2} - \dfrac{(c_{13}+c_{44})^2}{\left(c_{44}-c_{66} - \dfrac{c_{11}+c_{12}}{2}\right)}}{2\left(c_{66} + \dfrac{c_{11}+c_{12}}{2}\right)} \cdot \theta^2 \Bigg\}, \qquad (4.10)
$$

$$
v_d \simeq v_{d_0} \left[1 + \frac{\left(c_{44} - \dfrac{c_{11}-c_{12}}{2}\right)}{(c_{11}-c_{12})} \cdot \Theta^2 + \frac{2(c_{11}+c_{12})\left(c_{66} - \dfrac{c_{11}-c_{12}}{2}\right)}{(c_{11}-c_{12})(c_{12}+c_{66})} \cdot \phi^2 \right], \qquad (4.11)
$$

where $v_{i_0} = \sqrt{\dfrac{c_{66} + \dfrac{c_{11}+c_{12}}{2}}{\rho}}$, $v_{d_0} = \sqrt{\dfrac{c_{11}-c_{12}}{2\rho}}$ are phase velocities for longitudinal v_{i_0} and slow v_{d_0} of shear elastic waves along the [110] direction.

Taking into account the smallness of the angles φ, θ, γ, α, ϕ, Θ and taking into account the values of the components of the tensor

c_{ij} [184], and also setting tg $\beta_0 = v_{i_0}/v_{d_0}$ (we find $\beta_0 = 82.2°$), from expressions (4.7)–(4.9), taking into account (4.6), and also (4.10) and (4.11), we find expressions for the angles Θ and φ of the reflected elastic wave:

$$\Theta \approx 1.01 \cdot \alpha - 0.136 \cdot \theta + 0.137 \cdot (\varphi - \gamma) \cdot \alpha, \tag{4.12}$$

$$\phi \approx 1.02 \cdot \gamma - 0.02 \cdot \varphi + 0.83 \cdot \alpha^2 - 0.24 \cdot (\alpha + \theta) \cdot \theta. \tag{4.13}$$

It can be seen from the expressions (4.12), (4.13) that the angles Θ and φ that determine the orientation of the reflected transformed elastic wave are mainly determined by the orientation of the reflecting plane and are weakly dependent on the direction of the incident longitudinal elastic wave near the [**110**] axis of the TeO$_2$ crystal. Thus, the requirements imposed on the orientation of the (**110**) plane of the crystal on which the piezoelectric transducer is welded, as well as the parallelity of the intermediate bonding layers, are not as stringent as when the slow shear wave is directly excited in the AOM sound line. At the same time, the requirements for the accuracy of the orientation of the ultrasound-reflecting vertex face are quite high and practically coincide with the requirements for the accuracy of the surface orientation of the sample of the TeO$_2$ crystal on which the piezoelectric transducer of the transverse elastic waves is welded in a modulator of the traditional design. For the indicated surface, the maximum error in the orientation, according to the data of [124], should not exceed ±2′.

The advantage of the AOM design in question is the possibility of carrying out a purposeful correction of the propagation direction of the transformed elastic wave. If, as a result of observing the pattern of the ultrasonic field in transmitted light, the direction of propagation of the energy flux of the reflected elastic wave makes an angle $\tilde{\omega}$ with the [$\bar{1}$10] direction in the plane (001), which is the plane of the optical aperture of the device, then the wave vector of the elastic wave makes an angle ϕ with this direction, The angles ϕ and $\tilde{\omega}$ are related by (4.4). When correcting, the angle $\tilde{\omega}$ should be changed so that the direction of the energy flow of the elastic wave coincides with the direction of the length of the optical aperture modulator. It can be seen from expression (4.11) that this can be done by changing the orientation of the reflecting face of the sound transmission line to the angles γ and α by its polishing. Moreover, as the angle γ changes (this angle changes in the same plane as the

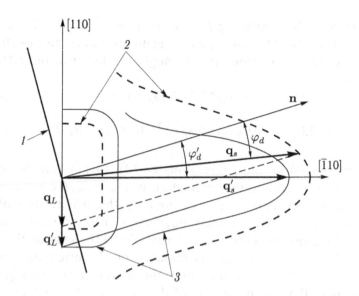

Fig. 4.18. Part of the section of the surfaces of the inverse phase velocities for longitudinal and slow transverse waves in the TeO_2 sound conductor by the **(001)** plane. *1* – projection of the reflecting face of the sound transmission line to the plane of incidence of the ultrasonic wave; *2* – sections of surfaces of inverse phase velocities for longitudinal and slow shear waves in TeO_2 at a temperature $T = T_0$; *3* – sections of surfaces of inverse phase velocities for longitudinal and slow transverse waves in TeO_2 at temperature $T_1 < T_0$.

angle ϕ), the angle ϕ varies rapidly, and with a change in the angle α, the angle Θ changes rapidly (see expression (4.12))

Performing similar calculations for the device shown in Fig. 4.17, *b*, we find: $\beta'_0 = 73.8°$,

$$\Theta' \approx 1.04 \cdot \alpha' - 0.29 \cdot \theta' - 0.3 \cdot (\gamma' + \varphi') \cdot \alpha', \tag{4.14}$$

$$\phi' \approx 1.08 \cdot \gamma' - 0.08 \cdot \varphi' + 1.85 \cdot \alpha'^2 - 1.05 \cdot \alpha' \cdot \theta' - 0.17 \cdot \theta'^2. \tag{4.15}$$

Thus, in this modulator design, the angles Θ' and ϕ', which determine the direction of propagation of the reflected ultrasonic wave, can be corrected by changing the orientation of the reflecting face of the sound line.

It should be noted that in the acousto-optical modulators and deflectors of the structure under consideration, correction of the direction of propagation of the reflected ultrasonic wave can be

a

b

Fig. 4.19. Photographs of acousto-optical modulators and deflectors with the transformation of the types of elastic modes on the verge of the sound line.

performed by changing the temperature of the sound transmission material when it is heated.

Figure 4.18 schematically shows a section of the surface of inverse phase velocities for longitudinal and slow shear waves in the TeO_2 sound line by the **(001)** plane at the temperatures of the sound line $T = T_0$ (curves *2*) and $T = T_1 < T_0$ (curves *3*).

As can be seen from Fig. 4.18, when the temperature of the sound line material changes, the reverse phase velocities for longitudinal and slow transverse elastic waves change, but since the incident and reflected elastic waves satisfy the relation (4.6), then with an unchanged orientation of the reflecting plane and a constant angle of incidence, the reflection angle φ_d changes. The dependence of the rate of change of the reflection angle φ_d on the temperature has the form:

$$\frac{\partial \varphi_d}{\partial T} = (\alpha_d - \alpha_i) \cdot \text{tg} \varphi_d, \qquad (4.16)$$

where $\alpha_d = \dfrac{\partial v_d}{\partial T} \cdot \dfrac{1}{v_d}$, $\alpha_i = \dfrac{\partial v_i}{\partial T} \cdot \dfrac{1}{v_i}$ are the temperature coefficients of the change in the speed of elastic waves. For the reflection geometry shown in Fig. 4.16 a, $\alpha_d = 2.11 \cdot 10^{-4}$ (1/deg), $\alpha_i = -1.26 \cdot 10^{-4}$ (1/deg) [185], which corresponds to $\dfrac{\partial \varphi_d}{\partial T} = -0.16$ (min/deg).

The principle of correcting the propagation of the reflected elastic wave due to a change in the ambient temperature was used in the AOD shown in Fig. 4.16 a, equipped with a thermostatically controlled heater, which made it possible to vary the temperature of the material of the sound line. Correction of the direction of propagation of the reflected ultrasonic wave was made in this device by changing the temperature when observing the pattern of the ultrasonic field in transmitted light.

Figures 4.19 a and b are the photographs of acousto-optical modulators and deflectors of the structure discussed above, which make it possible to correct the direction of propagation of an ultrasonic wave. These devices, along with AOMs of traditional design, (Fig. 2.20) were used in a system for displaying and recording television information with a pulsed copper vapour laser.

The AOM bandwidth for the information display system was 60 MHz at half power level at a central frequency of 80 MHz, and the deflector was at 70 MHz (45÷115 MHz). Moreover, in the deflector, the angle α was chosen equal to 5°, and in the modulator $\alpha = 0$. For the supplied electric power $P_e = 0.5$ W, the average Bragg diffraction efficiency in the frequency band was 80% for AOM, and 60% for AOD.

The diffraction conditions in the acousto-optical devices depend on the wavelength of the light source. Obtaining a multicolour image with a single acousto-optical deflector is difficult because of the complexity of matching the bands of the acousto-optical interaction at different wavelengths, so when forming a colour image, the AOD must be placed in each of the RGB channels and the TV rasters are reduced on the screen or in the intermediate image plane.

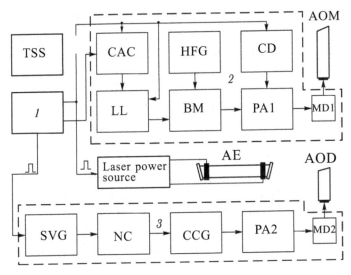

Fig. 4.20. Functional diagram of the device for controlling the acousto-optical television display system

4.4. Features of the operation of electronic control devices for the acousto-optical television information display system

Electronic devices that are part of the acousto-optical system for displaying television information with a pulsed copper vapour laser should perform the following functions:

1) providing synchronous operation of a pulsed laser in accordance with the parameters of a television scan;

2) generation of control signals for acousto-optical modulators and deflectors to ensure their effective operation;

3) conjugation of parameters of external information arriving at the system input, in accordance with the required control signals;

4) correction of the non-linearity of the amplitude transfer characteristic of the AOM;

5) ensuring the possibility of compensating for the attenuation of the ultrasonic signal along the length of the acousto-optical modulator;

6) ensuring the deviation of the light beam along the frame according to a linear law.

Figure 4.20 is a functional diagram of the device for controlling the acousto-optical television information display system, which includes: the interface unit (*1*), the system control unit in line (*2*), and the system control block by frame. The interface unit ensures the matching of the parameters of the signals coming from the television signal sensor (TSS) in accordance with the requirements of the normal operation of the units (*2*) and (*3*). From the incoming full television signal, the conjugation unit allocates a video signal, horizontal and vertical synchronization pulses.

Laser launch pulses are formed from the horizontal sync pulse. In the case of the laser generator–amplifier system, two sequences of trigger pulses are formed, shifted relative to each other by a value of $\tau_{???}$. The value of $\tau_{???}$ is adjusted during tuning of such a system at the maximum output power of the laser amplifier. Since the mode of coordinated operation of the generator–amplifier system is limited by the detuning time $\Delta\tau \simeq 25$ ns, the circuit must be able to adjust this value with an accuracy of not worse than $1\div2$ ns and maintain this value during operation.

The system control unit in line *2* serves to generate a radio signal with a carrier frequency f determined by choosing the operating point of the acousto-optical modulator (f', θ'_0, θ'_1) and the envelope corresponding to the input video signal. The modulation sign should be positive, the depth of modulation is 100%, and the power of the radio signal is such that an effective diffraction of light in the AOM at an allowable level of nonlinear distortion is ensured. In the developed system, this unit includes the following devices: the level lock (LL), the amplitude characteristic correction of the AOM (ACC), the high frequency generator (HFG), the balanced modulator (BM), the attenuation attenuator of the ultrasonic signal (AA), power amplifier (PA 1) and a matching device (MA1).

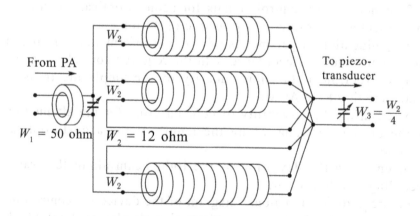

Fig. 4.21. Scheme of the device for coordination of AOM based on BTL.

Fig. 4.22. Photograph of the matching device together with the AOM,

The broadband balanced modulator is used for the amplitude modulation of the carrier frequency signal f' from a continuous generator (HFG) by a video signal coming from the amplitude corrector circuit. To obtain a high contrast in the image, the depth of modulation in the output signal of the BM must be maintained at 100% in a broad band of base frequencies. The balanced modulator used in the system provides carrier frequency suppression at zero level of the input signal by 40 dB, in the band of modulating frequencies from 16 kHz to 30 MHz at $f' = 80$ MHz. The amplitude transfer characteristic of the BM has a linear law of variation up to amplitudes of the input signal of 1.2 V. The level clamp located at the input of the balanced modulator makes the level of the 'black' in the television signal binding to a certain constant voltage level relative to which the balance in the BM scheme occurs.

The radio signal from the output of the balanced modulator is fed to the power amplifier, which provides the signal amplification by $30 \div 40$ dB so that the maximum level of the average electric power of the high-frequency signal fed to the acousto-optical modulator from TeO_2 is $0.5 \div 0.7$ W. The amplifier (PA 1) has a passband from 20 to 250 MHz with a non-uniformity of the amplitude–frequency characteristic not exceeding 3 dB in the specified frequency band.

From the output of the power amplifier, the signal through a coaxial cable with a wave resistance of 50 Ω is fed through the matching device (MD1) to the acousto-optical modulator. MD1 is designed to match the input impedance of the modulator with the wave impedance of the cable in the radio-frequency band. The circuit of the matching device is determined by the specific type of the frequency response of the input impedance of the loaded piezotransducer and depends on many parameters, in particular, on the relationship between the acoustic impedances of the piezotransducer and the sound line in which the desired type of ultrasonic wave is excited. The measurements show that when the slow shear wave is directly excited in the [110] direction of the TeO_2 crystal by a piezoelectric transducer of the X-cut from $LiNbO_3$ with a thickness equal to half the wavelength of the excited oscillations at a frequency of 80 MHz, the equivalent active part of the input impedance of the AOM is, as a rule, $2 \div 4$ ohms (electrode width $4 \div 5$ mm, height 4 mm). In this case, it is possible to obtain a good matching of the load with the cable by means of a broadband long-line transformer (BLT) with a transformation ratio of 1:4. The design of this type of transformer and the method of its calculation are described in [186] Figure 4.21 shows the scheme and Fig. 4.22 a photograph of the matching device based on such a transformer made on four segments of a microstrip line of the type RP-12-5 with a wave impedance of 12 ohm with a length of 60 mm on which ferrite rings of the type 1000 HM–10 × 6 × 4 are placed. Between the RC-50 cable and the transformer there is a matching balanced line with a wave resistance of 50 ohms, the braiding of which is ferrite rings.

Compensation of the reactive components of the input impedance is accomplished by connecting small capacitors at the input and output of the BLT. Measurement of standing wave voltage factor (SWVF) of the AOM from TeO_2 with the piezoelectric transducer of the X-cut and matching device based on BLT has shown that for the frequency range $40 \div 120$ MHz SWVF varies within the limits of $1.18 \div 1.64$.

When using the design of an acousto-optical modulator with the transformation of elastic vibrations onto the ultrasound-reflecting side of the sound line (see Fig. 4.16), the active component of the input impedance of the piezoelectric transducer is much larger than that of the conventional modulator. So a series of measurements of the input impedance of the AOM with piezoelectric transducers of the $Y_{+36°}$-cut $LiNbO_3$ and the central frequency of 80 MHz of this design showed that the active component of the input impedance varies within the limits of $10 \div 20$ Ohms with the capacitive nature of the reactivity. In this case, satisfactory matching in the operating frequency range can be achieved in a simpler way, for example, by means of a quarter wave length of the long line at the central operating frequency whose wave resistance is $W = \sqrt{W_0 \cdot \mathrm{Re} Z_{tr}}$, where W_0 is the wave resistance of the cable, $\mathrm{Re}\, Z_{tr}$ is the active component of the input impedance of the AOM piezotransducer. Compensation of capacitive reactance in this case is carried out by including a small inductance between the piezoelectric transducer and the long line. The measurement of the SWVF showed that for a modulator of this design, with the specified matching method, it varies within the range $1.2 \div 1.55$ in the frequency band $56 \div 100$ MHz.

The amplitude response correction (ARC) included in the video signal path at the input of the unit (2) is a wideband video signal amplifier whose amplitude transfer characteristic $U_2 = f(U_1)$ has a dependence allowing, in some approximation, to correct the nonlinearity of the amplitude characteristic of the AOM. In this case, the amplitude characteristic AOM is the dependence of the light intensity in the signal of the image on the modulation index χ_0. The square of the modulation index can be considered proportional to the power of the electrical signal,

$$\chi_0^2 = \frac{2\pi^2 \cdot P_a \cdot M_2 \cdot L}{\lambda_0^2 \cdot h}, \tag{4.17}$$

$$P_a = 0.5 \cdot \rho \cdot v^3 \cdot (A_0 \cdot \xi^0)^2 \cdot L \cdot h, \tag{4.18}$$

$$M_2 = \frac{n_o^6 \cdot p_{\mathrm{eff}}^2}{\rho \cdot v^3}, \tag{4.19}$$

where L is the width, h is the height of the piezoelectric transducer, ρ is the density of the sound-conducting material, $A_0 \cdot \xi_0$ is the deformation of the medium, p_{eff} is the effective photoelastic constant.

The power, in turn, depends linearly on the square of the voltage amplitude at the input of the balanced modulator. In the approximation of the third order of interaction, the amplitude characteristic of the AOM can be represented as a function (see expressions $(3.133) \div (3.142)$):

$$I = A_1 \cdot U_2^2 \cdot (1 - A_2 \cdot U_2^2 + A_3 \cdot U_2^4) = F(U_2), \tag{4.20}$$

where A_1, A_2, A_3 are certain constants depending in particular on the choice of the operating point on the angular-frequency characteristic of the AOM (f', θ_0', θ_1') the interaction length L, and also on the modulation frequency of the input signal f_0; $U_2 = b_2 \cdot \chi_0$ is the voltage amplitude of the signal at the input of the BM, proportional to the modulation index χ_0. The coefficient b_2 for a particular device is subject to experimental determination. The calculated form of the dependence (4.20) should correspond to the graphs in Figs. 3.19–3.21, from which it can be seen that the shape of the amplitude characteristic of the AOM depends on the frequency of modulation of the video signal. To compensate for the non-linearity of this characteristic, it is necessary that the transfer characteristic of the corrector has the form: $U_2 = F^{-1}(U_1)$, where $F^{-1}(x)$ is the inverse function. It is difficult to provide accurate correction in a wide modulation frequency band for the AOM. It is possible, for example, to split the video signal path in the equalizer into several parallel frequency channels that cover the whole frequency spectrum of the input signal, each with its own law of correction of the nonlinearity of the transfer characteristic. Further, these signals are added together, and the phase delays in the passage of signals for each channel should be aligned. The total corrected video signal is fed to the input of the balanced modulator.

In an acousto-optical modulator from TeO_2, the situation becomes even more complicated due to the fact that when the slow shear elastic wave propagates along the [110] direction, it experiences strong attenuation, which leads to uneven distribution of its amplitude along the length of the sound line. As shown in [181], the attenuation coefficient for this type of ultrasonic wave is 290 $\dfrac{dB}{cm \cdot GHz^2}$ dB, which corresponds to 6.5 dB attenuation at a 35 mm sound line length. Correct work of the amplitude corrector should be performed taking into account the equalization of the amplitude of the ultrasonic wave along the length of the modulator. To this end, an attenuation compensator (AC) is input to the system control unit via a line, which serves to output a voltage that changes the gain of one of the cascades power amplifier in such a way that the amplitude of the sound oscillations remains constant along the length of the AOM. The compensator produces a periodic voltage that varies according to the law $U = U_0 \cdot e^{-\alpha_0' t}$ ($0 \leq t \leq 1$, $\alpha_0' \simeq 0.36$). Equalization of the ultrasonic wave amplitude is achieved by increasing the average electric power at the output of the amplifier PA1.

The installation uses an amplitude corrector which was designed as the amplifier with an adjustable shape of the amplitude characteristic,

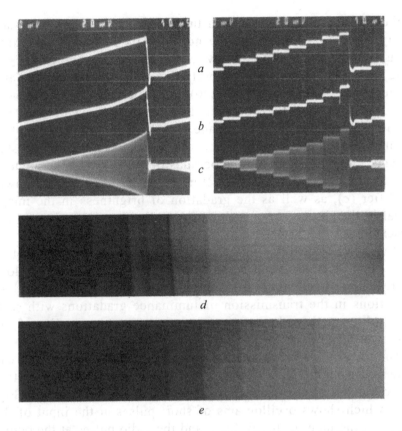

Fig. 4.23. Photos describing the operation of the electronic control system at low modulating frequencies.

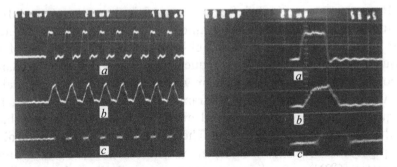

Fig. 4.24. Photos describing the operation of the electronic control system at high frequencies of the modulating signal.

analogous to the schemes of gamma correctors used in television [187], which had a passband of 50 Hz÷30 MHz. For standard video signals ($U_1 \leq 0.7$ V) coming from the television receiver, the shape of the amplitude characteristic of the equalizer was chosen in such a way as to compensate for the nonlinearity in the transmission of luminance gradations at low modulation frequencies ($f_0 \leq 1$ MHz). This characteristic was preliminarily selected on the basis of theoretical graphs (see Fig. 3.20) and experimental dependences (see Fig. 5.4), and then corrected by visual observation of the brightness distribution in the image signal when a signal is applied linearly to the input of the device varying voltage, as well as step voltage. Figure 4.23 shows photographs of the signal at the input of the corrector (*a*), after the corrector (*b*), at the input of the power amplifier (*c*), as well as the gradation of brightness in the image signal in the absence of correction (*d*) and with its presence (*e*) are presented.

It should be noted that in the visual perception of the light picture due to the physiological peculiarity of the human eye, the eye logs the real law of brightness distribution in the image field, therefore distortions in the transmission of luminance gradations with such a recording method do not significantly affect the quality of the perceived image.

The capabilities of the acousto-optical modulator control unit can be judged from the photographs in Fig. 4.23, reflecting the operation of the block at low frequencies, as well as from the photos in Fig. 4.24, which shows oscillograms of short pulses at the input of the ARC (*a*), the input of the BM (*b*), and the radio pulses at the output of the power amplifier (*c*). The duration of an individual pulse is approximately 50 ns.

If an acousto-optical deflector is used in the system to deflect modulated lines through the frame, the control unit for frame 3 (Fig. 4.20) serves to form the high-frequency control signal of the acousto-optical deflector. The frequency of this signal should change according to a law close to linear with a repetition period equal to the time of the frame scan (half-frame, $T_{fr} = 20$ ms) of the scan. The frequency range is chosen based on the operating frequency band of the acousto-optical deflector. The block consists of: a controlled chirp generator (CCG) or a frequency synthesizer, a sawtooth voltage generator (GPN), a nonlinearity compensator (KH), and a broadband power amplifier (PA 2) with a matching device (MD2). The sawtooth voltage generator has an external trigger from the frame sync pulses and produces a sawtooth control voltage of the chirp generator. The requirements imposed on the linearity of the vertical scanning of a light beam, depend on specific tasks that can be solved with the help of the system under consideration. The non-linearity of the vertical

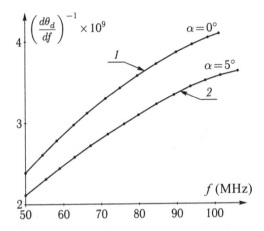

Fig. 4.25. The law of the change in the nonlinearity of the angular-frequency dependence of AOM (*1*) and ANM (*2*)

scanning can be estimated using the coefficient K_n [164], determined by the formula

$$K_n = 2 \cdot \frac{\left.\dfrac{d\theta_d}{df} \cdot f'\right|_{max} - \left.\dfrac{d\theta_d}{df} \cdot f'\right|_{min}}{\left.\dfrac{d\theta_d}{df} \cdot f'\right|_{max} + \left.\dfrac{d\theta_d}{df} \cdot f'\right|_{min}} \cdot 100\%, \qquad (4.21)$$

where $f' = \dfrac{df}{dt}$ is the rate of change of the frequency of the chirp signal, θ_d are the angles of the AOD deviation.

With the linear law of the change in the angles of the deviation of light from the frequency that occurs when isotropic light diffracts from the sound, the non-linearity of the sweep will be mainly determined by the nonlinearity of the chirp signal. In the case of anisotropic diffraction, the angular-frequency characteristic of the deflector is nonlinear. Figure 4.25 shows the calculated dependences of the inverse magnitudes of the diffraction angle change rates on

the frequency of the ultrasonic wave $\left(\dfrac{d\theta_d}{df}\right)^{-1} = \Psi(f)$ for acousto-

optical devices from TeO_2 for scattering geometries corresponding to the angles $\alpha = 0$ and $\alpha = 5°$. The first of these dependences ($\alpha = 0$) characterizes the non-linearity of the angular-frequency dependence of the acousto-optical modulator, and the second – the deflector,

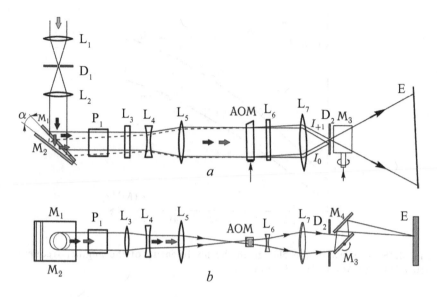

Fig. 4.26. Optical scheme of a projection device for displaying TV information in two mutually orthogonal planes (*a* and *b*). The upper drawing corresponds to a horizontal plane coinciding with the diffraction plane of the AOM.

which were used in the information display device. The calculations show that for $\dfrac{df}{dt} = \text{const}$ and frequency changes from 50 to 100 MHz, the value of the coefficient for the AOD from TeO_2 reaches 52%. This non-linearity can be compensated by the appropriate choice of the law of frequency variation of the chirp signal such that $\dfrac{df}{dt} = \text{const} \cdot \left(\dfrac{d\theta_d}{df} \right)^{-1}$.

For this purpose, a non-linearity compensator is added to the system control block along the frame, which corrects the shape of the sawtooth control voltage in accordance with curve *2* in Fig. 4.25.

4.5. Optimization of the optical scheme for the information display system with a pulsed copper vapour laser

In chapter 2 (see Fig. 2.22) we considered a possible variant of the optical scheme of the information display device with a pulsed laser, which was taken as a basis for a theoretical analysis of the imaging process using a pulsed laser. The advantage of this scheme is its simplicity and minimal requirements for the selection of optical elements. At the same time, the presented optical scheme has drawbacks, which limit the possibilities of efficient use of the copper vapour laser radiation. These disadvantages include the following technical problems.

First, the acousto-optical modulator, whose circuit is shown in Fig. 2.22, is located in the focal plane of the cylindrical lens. Therefore, when large laser radiation powers are used, the power density in the waist region reaches large values. This can lead to local heating and disruption of the optical homogeneity of the acousto-optical modulator's sound transmission material, leading to deterioration in the parameters of the projected television line, i.e., to reducing the number of solvable elements and limiting the size and brightness of the projected image. In the limit, at small values of the divergence of laser radiation, the sound line acoustic-optical modulator can be destroyed. In addition, the placement of the acousto-optic modulator sound line in the laser beam, coinciding with the object plane of the projection lens, leads to the fact that the projection system becomes extremely sensitive to various inhomogeneities of the refractive index of the material of the sound line and the modulator Windows, which are manifested in the form of vertical bands in the image..

Secondly, a copper vapour laser emits at two wavelengths $\lambda_1 = 510.6$ nm and $\lambda_2 = 578.2$ nm, whose average radiant power in the steady-state mode of heating the active element is approximately equal. When the AOM from TeO_2 is adjusted to the maximum diffraction efficiency for one of these wavelengths λ_1 at a selected ultrasonic frequency f and an angle of incidence θ'_{i1} conditions for effective diffraction at another wavelength, the incidence angle is not satisfied, or is performed for a portion of the modulation signal. As a result, when forming an image with a single AOM, it is necessary to use only one wavelength of the copper vapour laser radiation and thereby substantially reduce the efficiency of the use of light.

Thirdly, the instant of the appearance of the emission maximum of the spectral component λ_2 in a copper vapour laser lags, depending on the pumping conditions of the active element, for a time $\Delta\tau_0 \approx 10 \div 20$ ns with respect to the instant of the emission maximum of the spectral component λ_1. During this time, the ultrasonic wave in the AOM moves by the amount $\Delta l = \Delta\tau_0 \cdot v$. As a result, the light replicas of the TV signal line appearing at the AOM output for the light components with different wavelengths will be shifted relative to each other.

To eliminate these drawbacks, the optical scheme shown in Fig. 2.22 uses additional optical elements proposed in [188] and tested in the optical scheme of the information display device with a copper vapour laser. The essence of these changes is shown in Figs. 4.26 and 4.27.

In the optical scheme of the projection device shown in Fig. 4.27, a parallel light beam leaving the laser with linear polarization falls at an angle of $45°$ in the horizontal plane on the dichroic mirror M_1, which reflects the spectral component of the ray with wavelength λ_1 and passes λ_2. The spectral component of the laser radiation with λ_2 falls on the mirror M_2 at an angle of $45° - \alpha$ and after reflection passes through the mirror M_1 once more. Thus, the mirror system M_1 and M_2 splits the original light beam into two components with wavelengths λ_1 and λ_2 propagating at an angle 2α to each other. Further, both beams pass through a polarization transducer of laser radiation P_1, which changes their polarizations from linear to elliptical (close to circular) so that they are close to the proper polarizations of the light waves incident on the sound beam in the AOM crystal from TeO_2. P_1 is represented by a Fresnel rhombus with 30×30 mm windows enlightened for the spectral components λ_1 and λ_2 at a normal incidence angle.

The laser radiation transmitted through the polarization transducer P_1 is fed to the laser beam aperture transducer consisting of lenses L_3, L_4, L_5, which changes the dimensions of the parallel laser beams of the two spectral components in the horizontal plane to the sizes of the operating AOM equipment and forms converging beams in the vertical plane. After passing the aperture converter, the angle between the beam axes of the two spectral components λ_1 and λ_2 becomes $2\alpha/M$, where M is the magnification coefficient of the matching telescope from the lenses A_4, A_5. Both beams are incident on the acousto-optical modulator's sound conductor at angles θ_1 and θ_2 in the horizontal plane and then propagate inside the sound line at angles θ'_{i1} and θ'_{i2}. The arrangement of the mirrors M_1 and M_2 is chosen so that the laser beams with wavelengths λ_1 and λ_2 are aligned with each other and the AOM aperture in the diffraction plane. The solution of the system of equations (2.30) and (2.31) for one selected ultrasonic carrier frequency f and two wavelengths λ_1 and λ_2 makes it possible to determine θ'_{i1} and θ'_{i2}, as well as the external angles of incidence $\theta_1 = n_0 \cdot \theta'_{i1}$, $\theta_2 = n_0 \cdot \theta'_{i2}$ and the angle $\alpha = \dfrac{M \cdot (\theta_1 - \theta_2)}{2}$. At $f = 80$ MHz, $\theta_1 = 0.051$ rad, $\theta_2 = 0.049$ rad, $M = 1.6$, $\alpha \cong 1.6 \cdot 10^{-3}$ rad.

A spectrum of diffracted light waves is formed at the output of the AOM in the first order of diffraction corresponding to the spectrum of the amplitude-modulated sound signal. The L_7 lens displays the image corresponding to this spectrum in the plane of the screen (E). All the extra diffraction orders are filtered by diaphragm D_2. Since the source of diffraction of light waves is the same sound wave, its visualized images at wavelengths λ_1 and λ_2 must coincide spatially. For this, the optical system from the lens L_6 and the lens L_7 must be corrected for the absence of chromatic aberration.

The optical system in Fig. 4.26, in contrast to the optical system in Fig. 2.22 is constructed in such a way that the object plane of the image formed on the screen and the plane of focusing of the light beam after the aperture converter are spaced apart. This is ensured by the fact that the object plane of the anamorphic corrector consisting of a negative cylindrical lens L_6 and the projection lens L_7 coincides with the focal plane of the aperture converter, and the object plane of the projection lens with the output aperture AOM. As a result, there is no focusing of the light beam inside the AOM crystal. This makes it possible to work with large average powers of laser radiation.

In the optical scheme shown in Fig. 4.26, the focus plane of the aperture converter is located in front of the AOM. The installation of a cylindrical L_6 lens after the AOM makes it possible to focus the line on the screen in the vertical plane, while averaging the light field at the output aperture of the modulator vertically in the image plane, which reduces the effect of crystal defects and sound field inhomogeneities in the modulator on the image quality lines on the screen.

Structurally, the M_1 mirror is a flat glass plate with dimensions of $45 \times 30 \times 5$ mm with a thin-film dielectric coating on one of the large sides that passes the spectral components with λ_1 and λ_2 at an angle of incidence of 45°. On the opposite side of the plate a dielectric coating was applied reflecting the spectral component λ_1 and the transmitting spectral component λ_2 at an angle of incidence of 45°. Mirror M_2 is made on the same as mirror M_1, glass plate, but with reflective spectral component λ_2 coating at an angle of incidence of 45°.

As the converter of the apertures of the spectral components of the laser beam in the optical circuit of Fig. 4.26, the lenses A_3, A_4, and A_5, which were achromatized for wavelengths λ_1 and λ_2, were used. The cylindrical L_3 lens had a focal length of +30 cm. As L_4 and L_5, lenses with focal distances of −20 cm and +32 cm were used, which form a telescopic system with a linear magnification coefficient $M = 1.6$. The distance between the main planes of L_3 and L_4 was 10 cm. As a projection lens, L_7, a lens with a focal length of +30 cm and a working light aperture of 60 cm was used. The C_6 lens had a focal length of −20 cm.

To scan the image vertically in the optical circuit in Fig. 4.27, an electromagnetic mirror galvanometer M_3 with a deflecting mirror size of $30 \times 15 \times 1$ mm^3 was used with an angle of deviation of the light beam ±5° and a cut-off frequency of 800 Hz. The deflecting mirror of the galvanometer had a dielectric coating reflecting laser radiation with wavelengths λ_1 and λ_2 at an angle of about 45°.

To eliminate the third drawback associated with the temporary non-coincidence of pulses at different wavelengths after the AOM in the patent [188], it is proposed to establish a time shift compensator a television line for the radiation components λ_1 and λ_2, which combines the light replicas of the television line signal for the indicated components in the object plane of the L_7 lens.

Figure 4.27 shows the optical scheme of the compensator, the principle of which is based on the color separation of the image, the transfer of these images from the output aperture AOM (plane OP_1) along two optically identical channels and further aligning the images in the intermediate plane OP_2, coinciding with the objective plane of the lens L_7. To achieve these goals, the output aperture AOM (OP_1) is located before the separation mirror M_5 in such a way that it coincides with the focal plane of the lenses A_8 and A_9. The laser

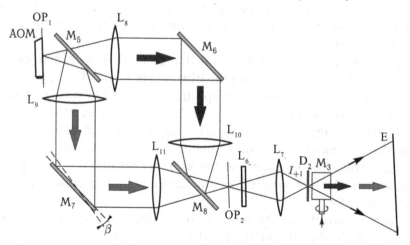

Fig. 4.27. Optical scheme of the time-shift compensator of images at different wavelengths.

radiation diffracted in the AOM, falling on the dichroic mirror M_5, is divided into two spectral components λ_1 and λ_2. The first spectral component passes through the mirror M_5, the lens A_8, and after reflecting from the mirror M_6, after passing through the lens A_{10}, after reflection from the mirror, M_8 coincides in the object plane of the projection lens L_7 with the second spectral component. The second spectral component of the image passes through its optical path of equal length and consisting of identical optical elements with the first optical path. By turning the mirror M_7 by a small angle β, it is possible to combine images on two spectral components in the plane OP_2. Next, this combined image is transferred by the L_7 lens to the screen.

Experimental investigations of the acousto-optical system of displaying and recording information with a copper vapour laser

5.1. Study of amplitude transfer characteristics and non-linear distortions in the formation of the image of a line

In chapter 3, a theoretical analysis of the process of image formation of a line with the aid of AOM from TeO_2 and a pulsed laser was carried out. As a result, an expression was obtained for the distribution of the light intensity in a row in the image plane, characterizing the response of the system to the input modulating signal of a harmonic shape. The calculated dependences presented in this chapter refer to an idealized system and are of an approximate, qualitative nature. At the same time, they make it possible to identify the main features inherent in the imaging system under consideration. So the above calculations show that when the amplitude of the input signal (modulation index) changes in the image signal, non-linear distortions arise, which are especially noticeable at low modulation frequencies.

An installation was assembled (Fig. 5.1) For experimental confirmation of this result and determination of the real amplitude transfer characteristics of the system during the formation of the image of the line. In this setup, a copper vapour laser with the active

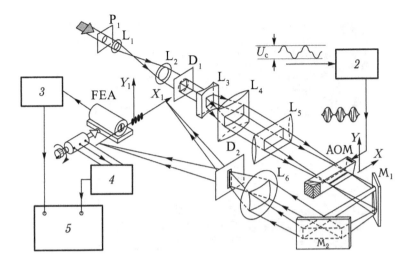

Fig. 5.1. Scheme of installation for the study of the amplitude transfer characteristics of the imaging system.

element Kulon-SM was used. The resonator of the laser oscillator corresponded to the optical scheme in Fig. 4.3. Linearly polarized radiation from this laser with generation parameters: $\lambda = 510.6$ nm, $\tau_0 = 7$ ns, $f_{rep} = 15.625$ kHz, passed through a quarter-wave plate P_1. After the plate, the polarization of the radiation became circular. The direction of rotation of the polarization vector was chosen in such a way as to obtain the regime of broadband anisotropic diffraction of light in the AOM.

After passing through the lens system $L_1 \div L_4$ and the aperture diaphragm D_1, a parallel light beam of rectangular cross section with dimensions of 4 cm along the X coordinate and 2 cm in Y was formed. With help a cylindrical L_5 lens with a focal length of 35 cm, the light beam was focused along the Y coordinate. The acousto-optical modulator from TeO_2 was placed in such a way that the light beam waist coincided with the plane of the output aperture of the modulator. The dimensions of the light spot along the X axis corresponded to the length of the AOM sound line. The width of the sound column in the AOM was 4 mm, and the carrier frequency $f' = 82.5$ MHz. A radio signal with an envelope varying according to the harmonic law and the modulation depth $m_0 \simeq 1$ was fed at the input of the modulator from the system control unit on line 2. At AOM, light was diffracted by an amplitude-modulated ultrasonic wave. With the aid of the L_6 lens, an enlarged image of the amplitude-modulated line on the screen was formed. To obtain a still image, the phase of

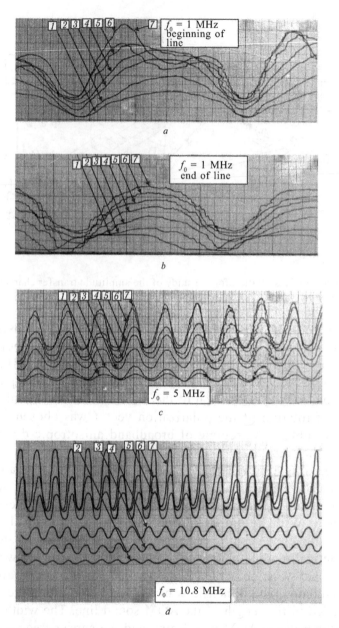

Fig. 5.2. Distributions of the light intensity in the image plane for the +1st diffraction order. The designations of the curves correspond to the following values of U_C: $U_C = 0.05$ V (1), 0.1 V (2), 0.2 V (3), 0.3 V (4), 0.4 V (5), 0.5 V (6), 0.7 V (7).

the signal was rigidly tied to the phae of the line pulses of a copper vapour laser. The diaphragm D_2, located in the plane of the spatial frequencies of the imaging lens L_6, selects the diffraction orders. In the image plane, a photomultiplier with a narrow slit was placed, by means of which the distribution of the light intensity along the line was recorded. The signal from the photomultiplier through the voltage follower *3* was applied to the input *Y* of the two-coordinate recorder *5*. The photomultiplier was moved along the *X* axis by means of a micrometric feed, and the unfolding voltage was removed from the multiturn potentiometer motor and fed through the voltage follower to the *X* input of the recorder. By moving the diaphragm D_2, it was possible to obtain an image in the +1st and 0th diffraction orders. At a constant power of laser radiation, the amplitude and the frequency of the input modulating signal changed. The depth of the modulation of the radio signal at the input of the AOM was maintained in all cases at level 1. The tuning of the modulator was carried out in such a way that the Bragg conditions for the angles of incidence and diffraction of light were satisfied for the central frequency $f' = 82.5$ MHz of the ultrasonic signal.

Figure 5.2 shows the obtained light intensity distributions in the image plane for the +1st diffraction order for different amplitudes U_C of a harmonic signal at the input of a balanced modulator and modulation frequencies of 1 MHz (*a, b*), 5 MHz (*V*) and 10 MHz (*d*). The distributions in Fig. 5.2 *a, c, d* correspond to ultrasonic signals at the beginning of the AOM sound line, and in Fig. 5.2 *b* for $f_0 =$ 1 MHz – at the end of the sound line.

The form of the obtained distribution for small values of U_C approximately coincides with the dependence (3.110). Figure 5.2 shows that the signal becomes distorted with increasing the amplitude of the input signal at a low modulation frequency (*a*), starting from a certain level.

Distortion increases with increasing signal amplitude. The increase in the amplitude of the image signal is slowed down. At high modulation frequencies there is almost no distortion the form of a signal and the intensity at the maximum of the image signal increases. Comparing the distributions in Fig. 5.2 *a* and 5.2 *b*, we can say that the amplitudes of the signal at the beginning and end of the line due to attenuation of the ultrasonic wave differ from each other by approximately 2÷2.2 times.

The nature of the distortions in the image signal of the line can also be traced from the change in its energy spectrum by recording the distribution of the light intensity in the plane of the spatial frequencies of the optical system of the line image formation. This plane coincides with the plane of the diaphragm D_2 in Fig. 5.1. In

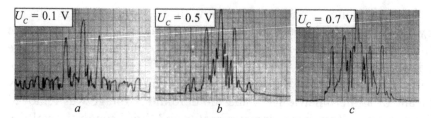

Fig. 5.3. The change in the energy spectrum of the image signal with a modulation frequency $f_0 = 1$ MHz with increasing amplitude U_C.

the experiment, a photomultiplier was placed in this plane, and on the output aperture of the modulator, a slit diaphragm was placed at the beginning of the sound line, the width of which was equal to several periods of the amplitude-modulated ultrasonic signal. This eliminated the ambiguity in the determination of the amplitude of the signal, which arises because of the attenuation of ultrasound. Figure 5.3 shows the change of the energy spectrum of the image signal with a modulation frequency $f_0 = 1$ MHz with increasing amplitude U_C. It can be seen from the figures that as the amplitude U_C increases, the spectrum of the image signal changes. Additional components appear in it, due to the non-linearity of the process of diffraction of light by sound. With increasing amplitude of the ultrasonic signal an increasing proportion of the light energy is concentrated in these components. In addition, the symmetry of the spectrum with respect to the central component is disturbed.

Figure 5.4 shows the experimental dependences of the relative light intensity at the maximum of the image signal for the +1st diffraction order on the amplitude of the input harmonic signal at various modulation frequencies, representing the real amplitude transfer characteristics of the acousto-optical line image forming system.

Comparison of these dependences with the calculated curves shown in Fig. 3.19–3.21, indicates that at high modulation frequencies ($f_0 = 5$ MHz and $f_0 = 10.8$ MHz), the experimental results agree fairly well with the theory for large and average values of the signal amplitudes U_C (we assume that $U_C = b_2 \cdot \chi_0$, where b_2 is a certain constant value). Some difference in the behaviour of the experimental curves from the calculated ones at small amplitudes can be explained by a decrease in the accuracy of measurements due to a large level of parasitic background illumination, as well as a decrease in the contrast in the input control radio signal.

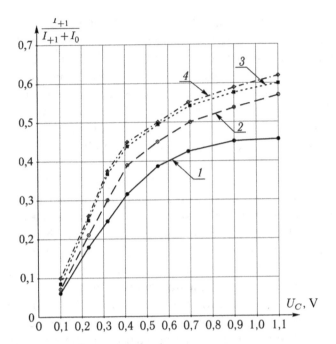

Fig. 5.4. The experimental dependences of the relative light intensity at the maximum of the image signal on the amplitude U_C of the input harmonic signal with the carrier frequency $f = 82.5$ MHz and the different frequencies of modulation of ultrasound: $f_0 = 1$ MHz (*1*), 3 MHz (*2*), 5 MHz (*3*), 10 MHz (*4*).

Measurement of the magnitude of the absolute contrast $K = I_{max}/I_{min}$ for the image signals shown in Fig. 5.2, showed that the maximum contrast, equal to $10 \div 15$, have signals at low modulation frequencies ($f_0 = 1$ MHz) and small signal amplitudes. When the amplitude of the signal increases at a low modulation frequency, the contrast decreases, which is due to the appearance of additional components in its spectrum at multiple modulation frequencies and a change in the energy balance between the central and lateral spectral components. As the modulation frequency increases, the contrast in the image decreases, which corresponds to the theoretical conclusions of chapter 2.

As the experiment showed, in a real system, the contrast in the transmission of amplitude-modulated signals falls in comparison with the theoretical one because of the large background illumination level due to multiple reflections and light scattering on the faces of the AOM crystal and the optical elements of the imaging system. A particularly large contribution to the background level is provided by the scattering of light on the output face of the AOM sound line, the plane of which coincides or is close to the objective plane of the lens

L_6 that is building the image. In the experimental setup (see Fig. 5.1), the optical surfaces of the lenses were not 'enlightened???' by the wavelength of the laser radiation, and the reflection of light from the faces of the acoustic waveguide with dielectric coating was 5÷8%. In addition, in a real system, the contrast depended on the depth of modulation in the radio signal at the AOM input. It was determined by the level of suppression of the carrier frequency at the minimum of the signal at the input of the balanced modulator of the BM and became less than 1 with increasing modulation frequency f_0 and a decrease in the amplitude of the modulating signal. The shape of the image signal becomes close to the harmonic law, and the contrast in the image decreases, which also corresponds to the theory.

Based on the experimental results obtained in the study of the amplitude transfer characteristics of an acousto-optical imaging system, a line with AOM from TeO_2 and a pulsed copper vapour laser, one can draw the following conclusions.

1. Theoretical consideration of the process of forming the image of a line, carried out in chapter 3, on the whole, it makes it possible to correctly describe the main regularities in the operation of the system with an input harmonic signal. The following experimental results agree with the conclusions of the theory:

a) the shape of the image signal,

b) patterns in the variation of the spectrum of this signal,

c) changing the contrast in the image at low modulation frequencies,

d) change in contrast with increasing modulation frequency,

e) regularities in the change in the amplitude transfer characteristics of the system, and, at large modulation frequencies, the numerical values of the diffraction efficiency of light at the maxima of the image signal approximately coincide with the calculated

2. To increase the contrast in the transmission of amplitude-modulated signals in the system under consideration, it is necessary to eliminate the scattering of light by the optical elements that form the image.

5.2. Experimental study of the characteristics of the acousto-optical television information display system on the projection screen

In order to determine the main characteristics of the acousto-optical information display system using a pulsed laser, a laboratory setup was created that makes it possible to form a television image on the projection screen. Such an installation, in addition to demonstrating the very possibility of a pulsed image formation method, allows us to evaluate the quality of the image directly, by visual perception, as well as to determine the main characteristics of the system: the number of solvable elements in a row and a frame, the linearity of the raster being formed, the possibility of transferring luminance gradations, the uniformity of the distribution of the light field by line and frame.

5.2.1. Optical scheme of the installation

Figure 5.5 shows the optical scheme of a laboratory installation for obtaining a TV image on a projection screen.

A copper vapour laser with the active element UL-101 was used as a light source. The optical scheme of the laser resonator corresponded to the scheme in Fig. 4.3. The unstable telescopic resonator was formed by spherical aluminum mirrors with radii of curvature of 2 m and 7 cm. The linear polarization of the radiation was set by a stack of quartz plates (8 pieces) located at the Brewster angle to the resonator axis between the AE and the mirror M_2. In order to eliminate astigmatism in the output laser beam, the stack was divided into two equal sections, rotated in different directions at the Brewster angle to the optical axis of the resonator. The average

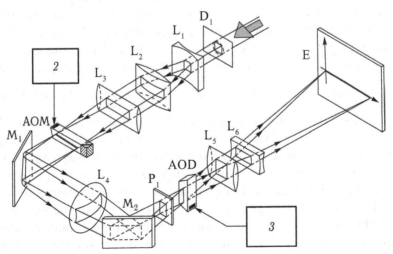

Fig. 5.5. Optical scheme of a laboratory installation for obtaining a TV image on a projection screen (E).

generation power of such a laser at a repetition rate of 15.625 kHz and an angular divergence of radiation of $3 \cdot 10^{-4}$ rad was 3.5 W, with approximately 30% of it occurring at a wavelength of 578.2 nm. The duration of the generation pulse was 10 ns at half-power level.

In order to increase the intensity of the diffracted light in the +1st diffraction order, the linearly polarized radiation from the laser was pre-transformed into an elliptically polarized radiation with the ellipticity and the direction of rotation of the vector of electric field polarization corresponding to the mode parameters of the light wave with the refractive index n_2 in the TeO_2 crystal (see expression (3.43)). This transformation was carried out using the Fresnel diamond (not shown in this figure). A light beam with a diameter $a_0 = 2$ cm from the laser entered the input of the optical system. The purpose of the optical elements A_1, A_1–A_4 of this system is the same as that of the elements A_1, A_3–A_6 in the circuit of Fig. 5.1. There was no need for a telescope from lenses L_1, L_2, such as that used in the circuit in Fig. 5.1.

The deviation of the light beam along the frame in the considered installation was carried out with the aid of an acousto-optical deflector (AOD), which was located in the Fourier plane of the L_4 lens. For effective operation of the AOD, certain requirements are imposed on the parameters of the light beam (see Section 2.6). So, in the plane of deflection along the frame the beam of light incident on the deflector must be parallel and have dimensions corresponding to the dimensions of the sound field in the AOD crystal, and the polarization of the incident light must correspond to the acousto-optical interaction required for the chosen geometry. In order to form a parallel light beam, the cylindrical lens L_3 and the objective lens L_4 are arranged in such a way that they form a telescopic system with the magnification factor $M' = \dfrac{F_2}{F_1} = \dfrac{D}{a_0}$, where F_1 and F_2 are the focal lengths of the lenses L_3 and L_4, respectively, and D is the aperture size of the AOD in the scattering plane. The installation used a deflector made of TeO_2, the direction of propagation of the ultrasonic wave in which makes an angle $\alpha = 5°$ with the [110] axis in the scattering plane (110) (the deflector design corresponds to Fig. 4.16 a). With this geometry, for effective use of the energy of the light incident on the deflector, its polarization should be close to linear with the polarization axis lying in the scattering plane. Since after the acousto-optical modulator the diffracted light has a polarization close to circular, then, in order to transform it into a linear one after the lens L_4, a thin quarter-wave plate P_1 was placed.

Since the AOD is located in the Fourier plane of the lens L_4, the distribution of the light field at its aperture corresponds to the spectrum of the spatial frequencies of the amplitude-modulated light signal. For undistorted reproduction of the image on the screen, the deflector must equally effectively deflect the light beams related to the diffraction orders from which this image is formed. The working aperture of the AOD, determined by the width of the sound column H in the deflector, is simultaneously a spatial frequency filter and determines the bandwidth of the modulation signal of the imaging system. Non-working diffraction orders are filtered by an opaque screen (not shown in the diagram in Fig. 5.5) located in the same Fourier plane of the lens L_4. The width of the sound column H and the focal length of the lens F_2 are interconnected by the relation

$$H \simeq \Delta\theta_{mod} \cdot F_2, \tag{5.1}$$

where $\Delta\theta_{mod}$ is the range of variation of the diffraction angles after AOM corresponding to twice the frequency band Δf_0 of the controlling ultrasonic signal. The quantity $\Delta\theta_{mod}$ as a function of f, f_0 and β is found from the solution of the system of equations (3.60) with allowance for (3.57). In the used deflector made from TeO$_2$, the size of the working aperture $(D \times H)$ was 15×10 mm. A TAIR-3-4.5/300 lens with a focal length of 300 mm was used as L_4, which allowed a modulated light beam with a modulation frequency bandwidth of up to 13 MHz to pass through the deflector. The focal length of the lens L_3 was equal to 400 mm, while the size of the light beam incident on the AOD in the plane of light deflection along the frame was 15 mm. Note that at such a focal length, the angle $\tilde{\alpha} = \dfrac{a_0}{2F_1 \cdot n_0}$, which characterizes the cone of the angles of incidence of light to the sound beam in the AOM, is $1.1 \cdot 10^{-2}$ rad and has practically no effect on the diffraction efficiency of light in the modulator.

The cylindrical lenses L_5 and L_6 after the AOD form a lens optical system having an angular magnification M_2' and tuned so as to focus the deflected light beams into the plane of the screen E (Fig. 5.5). The coefficient of the angular magnification M_2' was chosen so as to match the size of the TV screen on the screen in accordance with the accepted standard (l_y/l_x = 3/4). The value of the coefficient M_2' can be determined from expression

$$M_2' \simeq \frac{3 \cdot M_1' \cdot L}{4 \cdot \Delta\varphi_{sk} \cdot F_2 \cdot (M_1 - 2)},$$ (5.2)

where $M_1' = l_x/L$ is the coefficient of the system magnification with respect to the x coordinate, $\Delta\varphi_{sk}$ is the range of the angles change after the AOD. For $M_2' = 30$ and $\Delta\varphi_{sk} \simeq 4.3 \cdot 10^{-2}$ rad, which corresponds to $\Delta f_{chip} = 50$ MHz for the AOD made from TeO_2, $M_2' \simeq 2$.

The optical system was adjusted to a sharp image either by moving the AOM a little along the axis of the system, or by moving the lens L_4 along the same axis. In the second case, a slight deterioration in the parallelism of the light beam incident on the AOD practically did not affect the diffraction efficiency.

5.2.2. Experimental results

The first experimental results of the study of an acousto-optical television system with a pulsed copper vapour laser in Russia were obtained in [189–192]. In the experiments, the laboratory setup shown in Fig. 5.5 was used and the following tasks were solved. First, it was a demonstration of the very possibility of a system with a pulsed method of projecting a TV-line over a long period of time. Secondly, the problem of determining and adjusting its parameters such as the number of solvable elements in a row and a frame, the linearity of the raster being formed, the possibility of transmission of luminance gradations, the uniformity of the distribution of the light field along the line and the frame were solved. Thirdly, the possibility of an undistorted display of information coming from the 'air' via a television communication channel was demonstrated.

The acousto-optical system was controlled by means of electronic devices the work of which was considered in Sec. 4.4. To adjust the optical elements of the system, as well as to check the linearity and the number of solvable elements in a row, a control signal was sent to the input of the system control device in line 2 in the form of packets (8 pulses per pack). The pulse frequency in the packet was 10 MHz, and the duration of the individual pulse was $\simeq$30 ns.

Figure 5.6 shows a photo of radio pulses at the input of the AOM, and Fig. 5.7 the photographs of the raster obtained on the projection

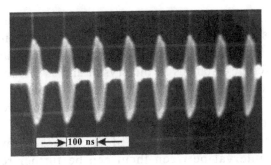

Fig. 5.6. Photograph of the radio pulses at entry to the AOM.

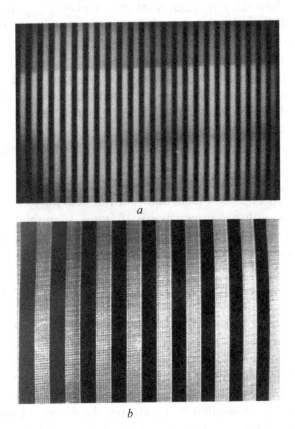

Fig. 5.7. Photos of TV images of bursts of short pulses.

screen, Fig. 5.7 *b* corresponds to the increased scale of the image in Fig. 5.7 *a*.

As can be seen from the photographs, the pulses in the packet are completely resolved, and the linearity of the signal sweep remains constant over the entire length lines. The observed small distortions at the edges of the resulting raster were determined by the aberrations of the optical system forming the image. These aberrations are mainly related to the fact that light beams falling at different angles on the AOD pass different distances in the crystal and as a result, the image of the extreme points of the line is located closer to the imaging optical system. The aberration data can be eliminated by appropriate correction of the optical system.

The high contrast in the image of the pulses allows us to say that the time interval between them can be further reduced. The number of solvable system elements per line is directly determined by the upper boundary frequency of the displayed signal f_{0b}, which depends on the width of the working aperture of the AOD and the focal length of the projection lens. In this case, it was significantly smaller than the limiting modulation frequency in the AOM. So for the given projection device $f_{0b} \simeq 13 \div 14$ MHz, in spite of the fact that the bandwidth of AOM together with its control device was equal to 30 MHz. The duration of the information part of the line T_c, which was used in this experiment, was 52 µs. According to formula (1.1), this corresponds to the maximum number of solvable elements in the line $N_c \simeq 640$ by the Rayleigh criterion, or 1280 by the TV standard, at $f_{0b} \simeq 14$ MHz.

The photo in Fig. 5.7 *b* also shows a line structure corresponding to 312.5 lines with a consecutive half-frame scanning, which was used in the experiment. Individual lines are fully resolved throughout the frame, and the distance between them allows one to place at least one more line.

To control the linearity of the scan over the frame, a test signal was applied to the acousto-optical modulator in the form of a sample of individual lines with amplitude modulation along the length of the line.

Figure 5.8 shows photographs of the control raster are presented. Figure 5.8 *b* corresponds to the increased scale of the image in Fig. 5.8 *a*. The number of lines in the frame was 312.5. Linearity of the scan by frame was estimated by measuring the distance between lines on the screen. The obtained value of the non-linearity coefficient of the scan (4.21) was approximately 10%.

To determine the capabilities of the laboratory installation for transmission of luminance gradations, as well as to adjust the amplitude response corrector (ARC), a linearly varying amplitude signal (10 steps along the line length) was fed to the system input

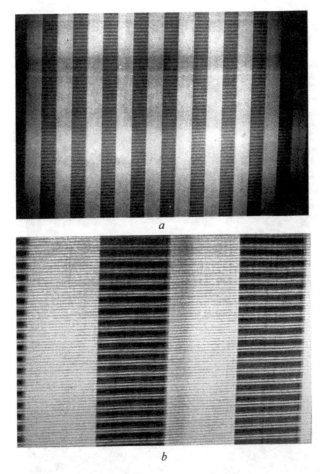

Fig. 5.8. Photos of the control raster of the linearity of the scan by frame.

(see Fig. 4.23). The obtained image of luminance gradations was presented earlier (see Fig. 4.23 *e*).

With the help of the installation in question, a television signal was displayed on a 1 m² screen. The signal was received from the aether both on the usual TV-receiver, and on the satellite communication system. A sheet of white paper served as a screen. Figure 5.9 shows pictures of the image of the television program, taken from the screen. The brightness of the resulting image was approximately 50 cd/m² at a wavelength of 510.6 nm.

In view of the fact that the laser radiation is coherent, a 'specular' structure was observed in the visual perception of the image, which led to a deterioration in the resolution at high frequencies. This structure could be destroyed by periodically moving the screen in the image plane. When watching a television signal, the line

Fig. 5.9. Photos of the image of the television program, taken from the screen,

structure was visible, which indicates the possibility of a laboratory installation to allow 625 lines per frame. The light losses in the experimental system were rather large and amounted to approximately 80–85% of the input laser radiation power, which is explained by the absence of antireflection coatings on the operating wavelength for optical elements and acousto-optical devices. In addition, the AOM efficiency at the maximum amplitude of the input signal at low modulation frequencies was not more than 50%, and the efficiency of the AOD used in the experiment was 60%.

The experimental results presented in this section were the first results in the USSR for obtaining a television image by the pulsed method of forming a line using a pulsed copper vapour laser and a fully acousto-optical light beam control system.

5.2.3. On the possibility of creating a multicolour projection device with metal vapour lasers

A colour laser projection device can be constructed by combining the three systems discussed above, the sources of radiation in which are copper vapour lasers (λ_{green} = 510.6 nm), gold (λ_{red} = 627.8 nm) and, for example, the second harmonic from a titanium-sapphire laser (λ_{blue} = 450 nm). In each of these three systems, a TV image is formed at its wavelength, and mixing is carried out by simultaneously overlapping them on a common screen.

On the other hand, as was said above (1.3), in lasers on a mixture of metal vapours simultaneous generation at several wavelengths on one active element is possible. For example, in a mixture of copper and gold vapour, we have a laser source at three wavelengths λ_{green} = 510.6 nm, λ_{yellow} = 578.2 nm, λ_{red} = 627.8 nm. In this case, a mixed image can be obtained with a single acousto-optical modulator per line. For this, three sound waves must be excited in the AOM. The frequencies of these waves, as well as the angles of incidence of light on the sound line, must be selected in accordance with the optimal conditions for acousto-optical interaction simultaneously for all light wavelengths. Since the sound beams are spatially coincident, it is easy to ensure spatial alignment of the television lines on the screen. To simplify the optical system after AOM, the ultrasound carrying frequencies and the angles of incidence of the light waves should be chosen in such a way that the diffraction angles for different carrier ultrasonic frequencies coincide. Calculations show that to satisfy this condition in a laser on a mixture of copper and gold vapour, the following incidence angles and ultrasonic carrier frequencies for different wavelengths can be chosen in the AOM from TeO_2: for different wavelengths: θ_i^{green} = 0.0218 (the angles of incidence are measured in the scattering plane from the axis [001] inside the TeO_2 crystal), f_{green} = 80 MHz; θ_i^{yellow} = 0.0196 rad, f_{yellow} = 64.2 MHz; θ_i^{red} = 0.0184 rad, f_{red} = 56 MHz. The light beams incident on the sound line for these wavelengths must be previously separated, so that their angles of incidence θ_i within the crystal take the values indicated above.

With a wide band of acousto-optical interaction, each of the incident light waves can diffract on all three carrier sound frequencies, and with a broadband signal modulation, respectively, on all three bands. For an undistorted information display with simultaneous excitation of three ultrasonic waves in an AOM crystal, the bands of the modulating frequencies must not overlap. So for displaying information simultaneously on the red and green lines at the above frequencies, the modulating signal band Δf_0 for these lines should not exceed the value $\Delta f_1 \simeq \dfrac{f_{green} - f_{red}}{2} \simeq 12$ MHz. If there is a blue line in the light beam, for example with $\lambda_{blue} = 450$ nm, then, in addition, the ratio $\Delta f_{blue} \leqslant \dfrac{f_{blue} - f_{green}}{2}$ must be satisfied. 'Parasitic' diffraction orders should be filtered out by a diaphragm located in the plane of the spatial frequencies of the imaging objective. To increase the band of operating frequencies while displaying information at several wavelengths with the help of a single AOM, it is necessary to move to the region of higher carrier frequencies of ultrasound. If subcarriers of sound frequencies fill the AOM in turn, then, naturally, there is no restriction on the band of modulating frequencies.

5.3. Application of an acousto-optical system with a pulsed copper vapour laser to record information

5.3.1. Recording of information on the space–time light modulator PRIZ

One possible application of an acousto-optical imaging system with a pulsed laser can be the recording of a two-dimensional array of data on optically controlled transparencies that are used in optical processing systems operating in real time. Such transponders, in particular, include the space-time light modulator (STLM) of the PRIZ type [193]. One of the main advantages of modulators of this type are relatively high noise characteristics, which makes it possible in principle to form large amounts of information on it. Nevertheless, the sensitivity of these STLM data at relatively high spatial frequencies ($20 \div 40$ mm^{-1}) is of the order of 50 μJ/cm^2, which requires high brightness from imaging devices. The generally used high-resolution projection television tubes in some cases do not provide the necessary exposures in the plane of the modulator, as well as high linearity of the raster being formed. This limits the practical capabilities of information processing systems that use STLM of the PRIZ type and makes it necessary to look for alternative imaging methods for this type of modulator. In works

[194, 195], the possibility of using an acousto-optical system with a pulsed method of forming an image of a string for recording information on the STLM PRIZ was tested, and also features of this recording mode were revealed. An installation was assembled that made it possible to form a small-sized television image on the surface of the STLM, corresponding to the diameter of its working aperture. Figure 5.10 shows the optical scheme of the installation and Fig. 5.11 the photograph of the installation itself.

The source of the recording radiation was a copper vapour laser with an active element Kulon-SM, which was considered earlier in section 4.1. The output parameters of the laser source were as follows: $\lambda = 510.6$ nm, $\tau_0 = 7$ ns, $f_{rep} = 15.625$ kHz, $P_{rad} = 200 \div 400$ W. The formation of a narrow converging light beam of the required polarization incident on the AOM in this system occurred in a manner analogous to that shown in the scheme of the previously considered measurement setup given earlier in Fig. 5.1.

After diffraction by an ultrasonic wave, the light beam entered the lens system A_6–A_9. With the help of the lenses L_6–L_8, the distribution of the light field in the horizontal plane from the output aperture of the AOM was transferred with a corresponding change in scale to the surface of the STLM PRIZ in the form of a thin line, modulated by the intensity of the video signal input to the system control device via the line.

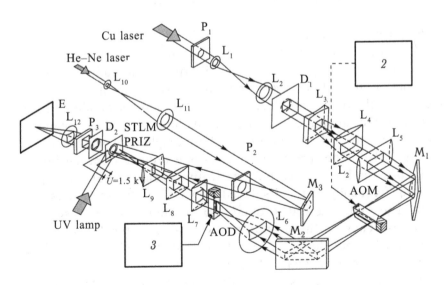

Fig. 5.10. The optical scheme of the installation for recording information on the STLM PRIZ.

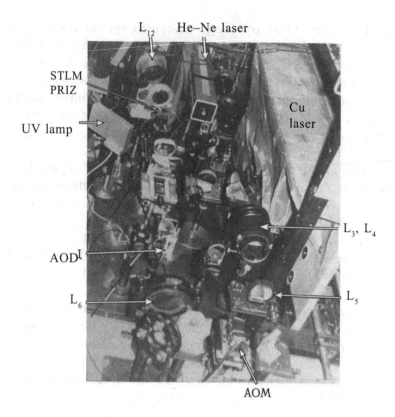

Fig. 5.11. Photo of the experimental setup,

Its position along the vertical coordinate was determined by the angle of the AOD deviation. To reduce the level of background illumination and to perform spatial filtering of the radiation in the constriction of the light beam along the Y axis, a slit diaphragm was located on the output aperture of the AOM, another diaphragm was placed on the input aperture of the AOD and selected the beams of the 0th and +1st diffraction orders (these diaphragms are not shown on Fig. 5.10). The resulting image was a rectangle measuring 14 mm along the line and 10 mm vertically (in the frame). The upper limiting frequency of the displayed signal f_{0b} in this setup was 14.5 MHz. The AOD allowed to obtain up to 1000 Rayleigh-resolvable positions. Thus, with an appropriate choice of the length of the television frame, the acousto-optical system made it possible to obtain a two-dimensional information array with a capacity of up to 900×1000 elements on the Rayleigh criterion at the working aperture of the PRIZ modulator.

As a modulator in the PRIZ system was a modulator from a $Bi_{12}SiO_{20}$ single crystal with a working aperture diameter of 16 mm, manufactured in the quantum electronics laboratory of the A.F. Ioffe Physico-Technical Institute. Using the STLM, the accumulation of lines and the subsequent coherent processing of the accumulated information array were carried out. The system in question worked in a cyclic mode: write–read–erase. After the voltage was applied to the STLM PRIZ ($U \simeq 1.5$ kV), one frame was recorded for 20 msec. Then the information was read out by a helium–neon laser beam with a wavelength of 632.8 nm. The modulator was located between the crossed polaroids P_2 and P_3. The spatial frequency spectrum of the diffracted light field was recorded in the focal plane of the L_{12} lens using a photoelectric multiplier. At the end of the reading, the voltage from the STLM was removed and the recorded information was erased by flash of the flash lamp, after which the STLM PRIZ was ready for the next recording cycle.

Studies were made for various video signals with 312 lines per frame. In this case, STLM was oriented in this way with respect to the polarization of the reading light in order to obtain the greatest diffraction efficiency on the spatial lattice recorded in the crystal. The energy of recording, taking into account the efficiency of the imaging system (2.5%), reached 120 µJ/cm². Figure 5.12 shows the spatial–frequency spectrum of the images of bursts of short pulses recorded on the STLM PRIZ (see Fig. 4.7). The image as a whole was an array of 1200 × 312 information samples.

The measurement of the absolute diffraction efficiency of STLM PRIZ for various spatial frequencies corresponding to different frequencies of the input modulating signal was made. The results of the measurements are shown in Table 5.1.

In measurements, an appreciable broadening of the diffraction orders with increasing spatial frequency was observed. So, for a spatial frequency of 35 mm⁻¹ (the distance between zero and diffraction orders is equal to 6.6 mm), the width of the diffraction

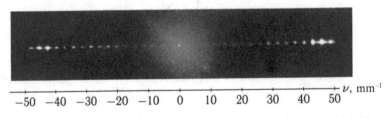

Fig. 5.12. Spatial-frequency spectrum of bursts of pulses, recorded on STLM PRIZ.

Table 5.1.

Spatial frequency, mm^{-1}	Information array (number of points by the Rayleigh criterion)	Diffraction efficiency,%	Energy recording, μJ/cm^2
20	600 × 312	1 × 10^{-3}	120
1	1000 × 312	2 × 10^{-4}	120
45	1200 × 312	0.5 × 10^{-4}	100

order was 40 μm, and for a frequency of 50 mm^{-1} (zero and diffraction orders are separated by 9.4 mm) this value reached 100 μm. The most significant reason for this broadening was apparently the aberrations in the optical system due to the thickness of the AOP sound line and, in addition, the insufficiently high quality of the cylindrical lens L_8 played a role.

The diffraction efficiency was measured for a different time of the write–read–erase cycle. It turned out that with a change in this parameter from 350 to 60 ms, the diffraction efficiency remained practically unchanged. But there was a change in the diffraction efficiency over time. The nature of this change is shown in Fig. 5.13.

It can be seen from the figure that, after 20–30 min of continuous operation (that is, after several hundred recording cycles), the STLM PRIZ practically loses its sensitivity. Apparently, this is due to the depletion of donor levels in a bismuth silicate crystal because of the high energy density of the recording light in some parts of the STLM surface. For example, with an available line width of not more than 10 μm, the average energy density per line is 400 μJ/cm^2, and in line sections where the signal is maximum, 800 μJ/cm^2.

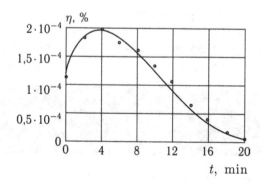

Fig. 5.13. Change in diffraction efficiency over time. The recording cycle frequency is 3 Hz, the spatial frequency is 35 mm^{-1}, the recording energy is 120 μJ/cm^2.

It took at least 1÷2 days to fully restore the sensitivity of STLM PRIZ. One of the possible ways to quickly restore the sensitivity is the illumination of STLM by infrared radiation. When the power of the infrared light source is several hundred milliwatts, several tens of seconds are enough to restore the sensitivity.

Based on the results obtained, it can be concluded that it is possible to construct information processing devices with a large information capacity on the basis of STLM of the PRIZ type and an acousto-optical information input system with a pulsed laser. As such, at the present time it is possible to use a solid-state pulsed laser with the transformation of radiation into the second harmonic.

The formation of an image of small dimensions (14×10 mm^2, the size of one element of 10×10 μm^2) at a high average power of light beams is of great interest also for technological purposes in problems of processing materials and marking them at high speed.

5.3.2. Recording television information on film

The simplest kind of medium on which it is possible to record television information with the help of the acousto-optical system under consideration is a film. Such a record in its time was useful for monitoring and long-term storage of information, for rewriting information from a VCR to film. Registration of the test signals generated by the image system on the film makes it possible to more accurately determine the distortions contained therein. Due to the fact that the laser light produced by the laser acousto-optical system contains a large amount of light energy, high-resolution fine-grained films (such as Mikrat) can be used to record information.

5.3.3. Recording television information on other carriers

The optical scheme of the installation, shown in Fig. 5.10, can be easily converted for the purpose of forming an image that is equal in size to that of a conventional photocamera (for example, 36×24 mm^2). In the setup considered in Fig. 5.10 to increase the image size on the line, the cylindrical lens L_8 with a focal length $F = 14$ cm was displaced along the Z axis in the AOM direction, and a cylindrical lens with $F = 50$ cm was used instead of the cylindrical lens L_9 ($F = 25$ cm) to increase the image size in the frame

In the image plane of the optical system, a camera was placed with a film on which a television raster was formed when the camera shutter was open. The shutter was opened from external synchronization frames for a time corresponding to several television frames. In order to adjust the number of frames (half-frames) in the image signal, a block was introduced into the electronic system management device, by means of which it was possible to adjust the number of synchronization pulses arriving at the start of the sawtooth voltage generator (SVG) of block 3 (see Fig. 4.20). In this case, the frame scan of the image was formed only when these pulses were received. The number of frames (half-frames) in the image signal could be changed from 1 to 8. The power of the light emission required for exposure was selected with the help of light filters located in the path of the laser beam to the optical system. For recording, we used a film of the Mikrat-300 type.

Figure 5.14 shows photographs of the resulting image of the test signals used to control the limiting number of the solvable elements of the system (*a, b*), as well as the linearity of the raster (*e, f*) being formed. The photos of the point field (Figs. 4.14 *c, d*) represent the response of the system to the input δ-shaped pulse signal (the pulse duration along the *X* axis was 50 ns, and along the *Y* axis the image dimensions corresponded to the width of the formed line) and actually characterized the optical transfer function of the investigated laboratory imaging system. Figures 5.14 *g, h* are photographs of a recorded television image corresponding to one complete television frame (625 lines). The input modulating signal of the acousto-optical system was a black-and-white video signal, which was extracted from the complete television signal coming from a TV receiver. Reception was carried out on the indoor antenna, so the signal contains distortions due to reception quality, as well as external interference. The line structure is clearly visible in the image. The quality of the displayed signal could be judged from the photograph of the television table (Fig. 5.14 *a*) recorded on the tape.

5.4. Displaying TV information on the big screen

To demonstrate the operation of the laser projection unit on a large screen in 1994, a portable system based on a copper vapour laser with an active element GL-201 (Kristall) was constructed, consisting of an emitter and an optical imaging system (in one block, Fig. 5.15, *a, b* on the color insert On 6), the modulator and high-frequency converter blocks (Figure 5.15, on the color insert On 6), and the system control devices (Figure 5.15, on the color. Insert 6). The laser

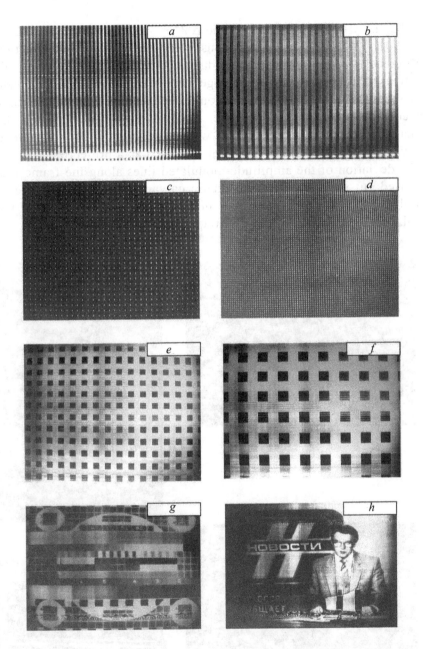

Fig. 5.14. Photographs of recorded images.

resonator circuit coincided with the circuit shown in Fig. 4.3, and the image was formed according to the scheme shown in Fig. 4.26. As already mentioned in Sec. 4.1, at a total light power of 22 W on two lines and at a pulse repetition rate of 15.625 kHz, the average power in beams with a good divergence $\Delta\Psi \simeq 1.6 \times 10^{-4}$ rad was 12 W ($P_{green}/P_{yellow} \simeq 1/1$) at a power consumption of 3 kW. The diffraction efficiency of the AOM at low modulating frequencies was $\simeq 40\%$. To increase the brightness of the image formed on the projection screen, two wavelengths of a copper vapour laser were used simultaneously according to the optical scheme shown in Fig. 4.26. To form the image on a remote screen, a lens with a focal length of +30 cm and a working light aperture of 60 cm was used as a projection lens L_7. The deviation of the amplitude-modulated lines along the frame was carried out using an electromagnetic galvanometer with a size of a deflecting mirror of $30 \times 15 \times 1$ mm^3, the angle of deviation of the light beam $\pm 5°$ and the cut-off frequency 800 Hz.

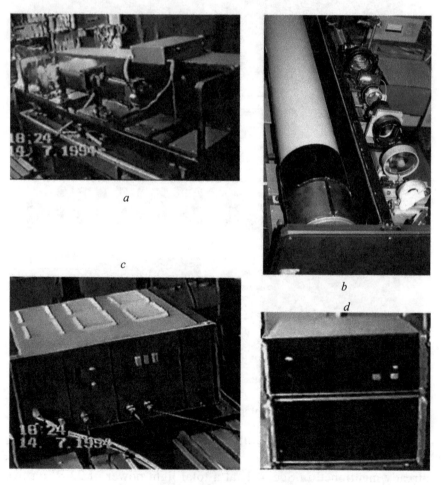

Fig. 5.15. Portable laser projection equipment.

Fig. 5.16. Operation of the laser p[rojection equipment.

Figure 5.16 presents photographs showing the operation of an acousto-optical projection installation with a copper vapour laser. Figures 5.16, *a, b* show the operation of the installation on the Lev Tolstoy Square in St. Petersburg in 1995. The distance from the installation to the screen was 80 m. The screen size consisting of rotary aluminum prisms with a reflective film deposited on their surface was 5 × 4 m. With this system, it was possible to display commercials pre-recorded on a video cassette, or an image coming directly from a television receiver.

Figures 5.16 *c, d* show the operation of the installation in a large concert hall 'Oktyabrsky'. In this case, the distance to the screen was 40 m, with an image size of 4 × 3 m. The laser projector replaced the projection equipment when projecting large-screen films from the VCR.

Figures 5.17 and 5.18 show the operation of a laser projection installation in laboratory conditions using a generator–amplifier with coupled resonators as a laser source, which was considered earlier in 4.1.3. Only one wavelength with $\lambda_1 = 510.6$ nm with an output power of 15 W was used to form the image.

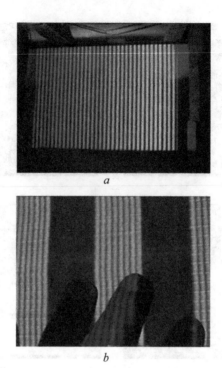

a

b

Fig. 5.17. Photograps of the image of the control raster.

The screen was located at a distance of 8 m from the projection lens with a focal length of 125 mm. The magnification of the image of line M_1 was 80. Figures 5.17 *a*, *b* show the screen images when the control raster is displayed in the form of bursts of short pulses of 50 ns duration (8 pulses per burst).

As can be seen from the photographs, individual pulses are completely resolved with high contrast along the entire length of the line (a total of 576 pulses or 1152 dark and light bands).

The image of the bursts of modulating pulses could be observed with reduced contrast when the screen was shifted relative to the position of the sharpest image within ± 4 m. The large depth of the sharp image along the X coordinate is due to the fact that when the image of the line is formed, the interaction of light with a sound wave modulated by the video signal occurs in the volume of the crystal for a sufficiently long interaction length $L = 4$ mm.

The depth of the sharp image of the amplitude modulated signal is $\Delta Z_1 \simeq \Delta L \cdot M_1^2$, where ΔL is the size of the sound beam along the Z coordinate, within which a line image is formed on the screen.

a

b

Fig. 5.18. Photographs of the image of TV transmission.

On the other hand, the longitudinal dimension of the focal spot responsible for the sharp image along the Y coordinate, according to [141], is $\Delta l = \pm 2 \cdot \lambda \cdot \left(\dfrac{F}{a_0} \right)^2$, where F is the focal length of the cylindrical lens L_5 (see Fig. 4.26), a_0 is the diameter of the light beam in accordance with the coordinate Y. At $F = 40$ cm, $a_0 = 2$ cm, the depth of the sharp image of the constriction is equal to $\Delta Z_2 \simeq \Delta L \cdot M_2^2 \simeq 2.6$ m. The large depth of the sharp image in two coordinates allows using vibrating screens to suppress the speckle structure using vibrating screens with an uneven surface, and also arrange the screen under a small angle to the imaging system and thereby reduce the transverse dimensions of the TV system. The resulting trapezoidal distortion of the raster and the nonlinearity of the scanning along the frame can be compensated electronically.

Figues 5.18 *a*, *b* show pictures of a television image taken from the screen of a laser TV projector.

5.5. Prospects for creating an acousto-optical system for displaying TV information in the high-definition standard with pulsed lasers

5.5.1. Study of the operation of the acousto-optical television information display system in the standard of enhanced definition

As a result of research into the operation of the laboratory acousto-optic television information display system, the optical scheme of which is shown in Fig. 5.5, the problem arose of investigating the possibility of its operation in the standard of high definition with a number of solvable elements in the frame of more than $10^3 \times 10^3$ by the Rayleigh criterion. One of the proposed for the implementation of standards for the decomposition of a television image was a standard having 1125 lines per frame at 25 frames per second. For this standard, the line length is 35.56 μs for the information part of the line 27.5 μs. The frequency of the horizontal sync pulses is 28.125 kHz.

For the qualitative reproduction of a television image in the specified standard, the following conditions must be met. First, the resolution of the system on the line should ensure the reproduction of single pulses of 21 ns duration (0.5 level), or have 1024 Rayleigh-solvable elements per line. It is seen from expression (1.1) that to obtain 1024 solvable elements in a row at T_c = 27.5 μs and Δf_0 = 50 MHz, it is necessary that $\tau_0 \simeq 7$ ns. Such a duration of a light pulse in a copper vapour laser at a wavelength of 510.6 nm can be obtained with a resonator length of $50 \div 70$ cm with the use of active elements of short length (AE type Kulon-CM, Kulon-M). To increase the output power of laser radiation, this radiation can later be passed through an amplifier based on a more powerful gas-discharge tube (such as Kristall).

Secondly, with a band of modulating frequencies of the video signal Δf_0 = 50 MHz, it is required to use the AOM with an acousto-optical interaction band $\Delta f = 2 \cdot \Delta f_0$ = 100 MHz. At T_c = 27.5 μs, this band can be provided in the AOM from TeO_2 with a length of a 17 mm sound line cut along the [110] direction of propagation of a slow shear wave and operating in a frequency band from 100 to 200 MHz with a carrier frequency of $f \simeq 150$ MHz.

At a really achievable value of the angular divergence of the laser radiation $\Delta\Psi_0 \simeq 10^{-4}$ rad to obtain 1125 solvable elements in the frame, the angle of deflection of the deflector $\Delta\varphi_{sk}$ should be $\simeq 7 \div 8°$. Such an angle is practically impossible to provide with the aid of an acousto-optical deflector. For scanning in a frame, in this case one needs to use either a polyhedral rotating drum, the construction of which is described in detail in [196], or an electromagnetic mirror galvanometer similar to the G 100 type optical scanner from General Scanning Inc.

The most problematic aspect in the proposed system variant is the development of an effective acousto-optical modulator with an acousto-optical interaction bandwidth up to 100 MHz, comparable in diffraction efficiency with the AOM from TeO_2, operating in the frequency range from 60 to 100 MHz. One of these modulators could be a modulator from TeO_2, in which the skew slices with $\alpha \simeq 5 \div 6°$ are used similarly to the deflector. In this case, the difficulty is caused by the fact that the wave and ray vectors of the elastic wave constitute a large angle between each other, and when using a conventional projection optical system, the image of the beginning and end of the line will be at a different distance from the lens, i.e., the image plane of the video signal will be rotated at some angle relative to the plane of the sharp image of the lines. In addition, the increase factor of the beginning and end of the line will be different. The use of AOMs from other materials to modulate a signal of this duration and bandwidth is problematic.

It seems that 'faster sections' of TeO_2, α-HJO_3, and $PbMoO_4$ crystals should be used as an AOM soundtrack for a system operating in the high definition standard. A very promising crystal for creating an AOM operating in a high-definition TV standard is calomel ($HgCl_2$). Thus, the speed of a longitudinal sound wave along the [001] direction in this crystal is $1.62 \cdot 10^3$ m/s, with a relatively small attenuation at a frequency of 200 MHz. The acoustic quality factor M_2 is more than 500 [197].

To operate in the specified TV-standard, the laboratory installation, assembled according to the scheme in Fig. 5.5, has been modernized. The active element in the laser was a sealed-off gas discharge tube of the Kulon-M type. A thyratron TGI-1-1000/25 was used for pumping. The discharge circuit for pumping the AE was made as in Fig. 4.1, with the values of the working and sharpening capacitances

$C_1 = C_2 = 550$ pF, $C_3 = 110$ pF. The AE was cooled with a thyratron and saturating chokes L_2 and L_3 using water. The laser resonator was assembled according to the scheme in Fig. 4.3, in which aluminum spherical mirrors with $R_1 = 4$ cm and $R_2 = 1.5$ m were used. Measurements were made of the duration of the laser pulse and the angular divergence of the laser radiation after the light beam selector (A_1, D, L_2). The measured radiation pulse duration at the half-power level was 7 ns, and the angular divergence was 10^{-4} rad, with an average radiation power of 2.5 W at a frequency of 28.125 kHz. The ratio between the powers of radiation on the green and yellow lines was 1/1.

To carry out the experiments in this TV standard, the AOM from TeO_2 was fabricated, the construction of which is shown in Fig. 4.15, and Fig. 5.19 shows a photograph of this AOM.

The length of the AOM soundtrack along the [$\bar{1}10$] direction was 20 mm, with 16.8 mm of this length coming from the information part of the TV line. The X-cut $LiNbO_3$ piezoelectric transducer was ground to a thickness of 20 μm, which corresponds to half the ultrasound wavelength at a central frequency of 120 MHz. To provide the acousto-optical interaction band at □ 80 MHz, the width of the deposited electrode was 2 mm. The maximum diffraction efficiency of this modulator in the frequency band 80 ÷ 160 MHz was approximately 30%. At these operating frequencies, the polarization of the incident and diffracted light waves is close to linear, so the linearly polarized laser radiation could be fed directly to the AOM without converting it to a circular one. The attenuation coefficient

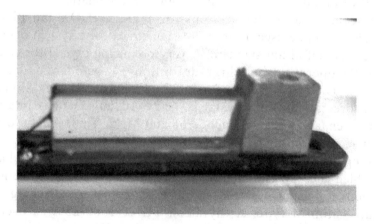

Fig. 5.19 Photographs of the AOM made from TeO_2 for operation in the HDTV standard.

for a slow shear wave propagating along the [110] direction in a
TeO_2 crystal is 290 dB/cm · (GHz)². This corresponds to a 7.1 dB
attenuation with a 1.7-cm soundtrack length at 120 MHz. In order
to equalize the amplitude of the ultrasonic wave along the length of
the AOM, as well as in the case of lower frequencies, an attenuator
was added to the control unit of the system, changing the gain of
the power amplifier as a function of time.

The deviation of the light beam along the frame in this experiment
was carried out with the aid of an electromagnetic galvanometer,
which was located in the Fourier plane of the lens L_4 in place of
the AOD. A diaphragm was placed in front of the mirror of the
galvanometer in the plane of the spatial frequencies of the image
forming optical system, which passed only the working diffraction
orders. A ready-made G 325 D Optical Scanner from General
Scanning Inc. was used as a galvanometer. The working aperture
of the galvanometer mirror in coordinate X is a filter of spatial
frequencies and, in the same way as the width of the sound column H
in the ADP, determines the bandwidth of the modulating frequencies
Δf_0 of the system in accordance with formula (5.1). The value of
$\Delta\theta_{mod}$ as a function of Δf_0 is found from the angular-frequency
characteristic for anisotropic diffraction of light in AOM from TeO_2.
At a frequency f = 120 MHz, we can assume that

$$\Delta\theta_{mod} \approx \frac{\lambda \cdot 2 \cdot \Delta f_0}{v}.$$ (5.3)

At λ = 510.6 nm, Δf_0 = 40 MHz, v = 0.61 · 10³ m/s, F_2 = 12.5 cm,
it follows from (5.1) that the required mirror size in the coordinate
X = (A) should be not less than 8.4 mm.

The size of the galvanometer mirror in the coordinate Y = (B) is
chosen from the following considerations. The number of solvable
system elements per frame is determined from the expression
$N_K = \frac{\Delta\varphi_{sc}}{\Delta\Psi_1}$, where $\Delta\varphi_{sc}$ is the scanning angle, $\Delta\Psi_1$ is the angular
divergence of the light beam at the output from the deflector, which
can be found from the relation $\Delta\Psi_1 = \Delta\Psi_0 \times \frac{a_0}{B \cdot \cos\varphi}$, where a_0 is
the diameter of the laser beam with angular divergence $\Delta\Psi_0$, φ
is the angle of incidence of the light beam on the mirror of the
galvanometer.

At $\Delta\Psi_0 \simeq 10^{-4}$, $a_0 = 12$ mm, $N_K = 10^3$, $\varphi = 45°$, we can find the relationship between the scanning angle of the galvanometer and the mirror size in the coordinate Y:

$$\Delta\varphi_{sc} = N_K \cdot \Delta\Psi_0 \cdot \frac{a_0}{B \cdot \cos\phi} \simeq \frac{1.7 \cdot 10^{-3}}{B}. \qquad (5.4)$$

At $B = 10$ mm, $\Delta\varphi_{sk} \simeq 10°$. The range of deviation angles of the galvanometer $\Delta\varphi_{sc}$ is a certain functional dependence on the time of the back stroke, the mass of the mirror and the size of the mirror along the Y coordinate. To reduce the retraction time, it is necessary to reduce the range of angles and dimensions of the mirror. The most suitable galvanometers for the vertical scanning, at present, are the galvanometers of the company General Scanning inc. G100, G300. To match the size of the TV screen on the screen by two coordinates in this system, you can change the range of deviation angles of the galvanometer.

The optical system is adjusted to a sharp image by moving the AOM along the axis of the system a little. By moving the L_4 lens along the same axis, the system is adjusted to a sharp image of the line along the Y coordinate.

The scheme of control devices for the laser projection equipment in the standard of high definition as a whole corresponded to Fig. 4.20. The initial control signals were the signals from the higher-resolution TV simulator, which consisted of different numbers and letters. To check the linearity and the number of solvable elements, a control signal was sent to the input of block 2 in the form of bursts of pulses (8 pulses per pack). The time interval between pulses in the packet was 50 ns, and the duration of an individual pulse was ≃20 ns. As shown by the visual observation of the image on the screen, the pulses in the packet were completely resolved, and the linearity of the signal scan remained constant throughout the entire length of the line.

5.5.2. Variants of creating a colour acousto-optical system with pulsed lasers operating in the HDTV standard

One of the most important tasks in the formation of a TV image approaching the quality of a photographic image is to enable the laser projection system to operate in a high-definition standard, for example, HDTV. For this standard, the number of solvable elements in the frame is 1080 × 1920 (9:16 format). The frequency of the change of frames is determined by the ability of the display system and should be 70÷100 fields per second.

Let us dwell, for example, on the simplest for the implementation of the interlaced standard for the decomposition of a television image with a frequency of changing the fields $f_f = 70$ Hz. The duration of the half-frame in this case is $T_{hf} = \dfrac{1}{f_{hf}} \cong 14.286$ ms ms, $T_{hf} = T_f + T_{rev}$, where T_{hf} is the duration of the working part of the half-frame, and T_{rev} is the duration of the reverse stroke of the vertical scan.

$T_{hf} = 540 \cdot f_c + \dfrac{n}{f_c}$, where f_c is the frequency of the line sync pulses, and n is the integer number of lines located on T_{rev}.

The duration of the reverse stroke of a vertical scan is determined by the speed of the electromagnetic galvanometer used in the system. We choose $n = 46$, then $f_c = 41.02$ kHz, and $T_{rev} \cong 1.1214$ ms, which corresponds to the speed, for example, of the galvanometer manufactured by Cambridge Technology Inc., model 6800/CB6588. The length of the line is $T_c = 1/f_c \cong 24.378$ μs. With the impulse method of forming a line image, there is no reverse horizontal scan, so the entire duration of the line can be used to display information on the screen, which also increases the resolution. This can be done with a digital processor that converts the incoming full-height TV signal into separate signals necessary for the functioning of the system.

We assume that the information duration of the line is $T_c = 24$ μs. The number of elements in a line according to the Rayleigh resolution criterion is $N_c = \dfrac{T_c}{\tau_0 + \tau_e}$, where τ_e is the duration of one element by the level of 0.5, and τ_0 is the duration of the light pulse. If we assume that the 1920 elements in the television standard correspond to 960 elements by the Rayleigh criterion, then $\tau_0 + \tau_e \cong 25$ ns. Thus, with a frequency band of the modulating signal $\Delta f_0 = 50$ MHz, $\tau_e \cong \dfrac{1}{\Delta f_0} = 20$ ns , therefore, the duration of the light pulse of the laser should not exceed 5 ns.

With a frequency band of the modulating signal $\Delta f_0 = 50$ MHz for amplitude modulation, it is necessary to use the AOM with an acousto-optical interaction band $\Delta f = 2\Delta f_0 \cong 100$ MHz. At $T_c = 24$ μs, such a band can be provided in the AOM from TeO_2 operating in the frequency band from 100 to 200 MHz with a 14.8 mm length of the sound line cut along the propagation direction of the slow shear wave [**110**]. The diffraction efficiency in such AOM will not be more than 30%.

To obtain a high diffraction efficiency with such a band of acousto-optical interaction, it seems that it is necessary either to use an AOM with a multielement piezoelectric transducer in which the Bragg angle is adjusted (which is difficult to realize in practice) or to find a way to adapt a modulator of TeO_2 using oblique ($5 \div 6°$) slices. In the latter case, as already noted in 5.5.1, the difficulty is caused by the fact that the wave vector and the direction of the energy flow of the elastic wave in the sound pipe constitute a large angle between each other. When using a conventional projection optical system, the images of the beginning and the end of the line will be at a different distance from the lens and have a different magnification. The use of AOM made from other materials to effectively modulate a signal of this duration and bandwidth is problematic.

In connection with all of the above, it seems promising the following option to create an acousto-optical system for displaying high-definition television information. The laser operates at a repetition frequency half as small as in the previous case ($\cong 20.5$ kHz), and for a single light pulse of 10 ns duration it displays two lines simultaneously on the screen. The system has a processor that converts lines in duration by a factor of two from 24 to 48 microseconds while preserving their information capacity. In this case, the frequency band Δf_0 should decrease from 50 to 25 MHz, and the acousto-optical interaction band should move to the region of the broadband interaction that is most effective for TeO_2 in the frequency range $50 \div 100$ MHz.

The system must simultaneously input two lines into two parallel modulators (one above the other) at a speed half that of the previous case. One can also use two-channel AOM from TeO_2. Such a line imaging system will save a large number of solvable elements with relatively low requirements for AOM and the duration of the laser pulse. A possible drawback of this method of displaying two or more lines per laser pulse is the reduction in the frequency of repetition of fields in a vertical scan, which can negatively affect the perception of the image by the eye.

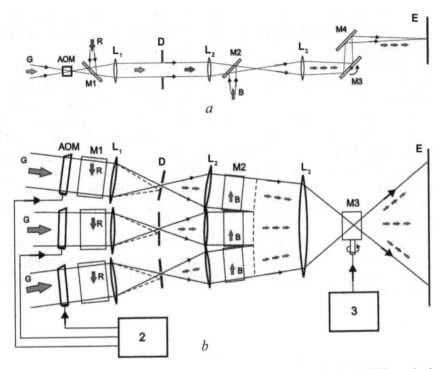

Fig. 5.20. Possible scheme of formation of the image of a line in the HDTV standard in two mutually orthogonal planes (*a*) and (*b*) using three acousto-optical modulators in each colour channel.

Another variant of an acousto-optical projection system with a pulsed laser can be a system in which a line image is constructed using two or more acousto-optical modulators arranged in series.

Figure 5.20 provides a possible scheme for forming a line in two mutually orthogonal planes using three acousto-optical modulators in each colour channel in the scanning plane along the frame (*a*) and the image plane of the line (*b*).

In this case, the length of the sound line of each modulator is 24 µs, and the number of elements solvable by each modulator is reduced by a factor of three (320 by the Rayleigh criterion). The electronic digital processor *1* (Fig. 5.20) splits the video signal into three parts, which are simultaneously input into the modulators placed consecutively one after the other. After all the sound lines are filled with an amplitude-modulated ultrasonic signal, the laser generates a short light pulse. Using the lens system L_1, L_2 and diaphragms D, the image from each modulator is transferred to the intermediate plane where images are added. Then, with the help of the common lens L_3, the total image is transferred to the screen E. In this case, the acousto-optical interaction band of the AOM can be no more than 40 MHz with a light pulse duration of 20 ns.

Channels of red R and blue B colours can be placed above and below the main green channel G. A colour mixing is performed using dichroic mirrors M_1 and M_2. The deviation of lines by frame is made with the help of a common galvanometer M_3. Thus, in this system, 9 acousto-optical modulators with a 15-mm sound line length should be used to form a full-colour wide-screen TV image. As already mentioned above, when using an AOD to deflect light on a frame, it is necessary to use separate AODs in each of the RGB channels, followed by the reduction of the rasters on the screen using a system of dichroic mirrors.

This formation scheme can be taken as a basis for designing ultra-wide-band acousto-optical TV image formation systems, in which a television line with a very large number of solvable elements can be obtained.

Conclusion

Acousto-optical systems for the formation of television images can be used to control the emission of powerful pulsed lasers (including infrared ones), as well as in special systems requiring, for example, the absence of mechanical deflection systems for the laser beam; in laser systems underwater location to create a synchronous illumination of underwater objects.

The results of the theoretical studies carried out in this book are of a general nature and can be used to find diffracted light fields in the intermediate regime for arbitrary geometries of acousto-optical interaction for acoustic crystals possessing, in particular, gyrotropic properties. These results allow to calculate the diffracted light fields in the intermediate diffraction regime for the complex frequency spectrum of the ultrasonic signal and for its amplitude distribution along two coordinates in the sound line.

The results of a theoretical analysis of the diffraction of light on an amplitude-modulated ultrasonic signal are also valid for a continuous laser and a laser operating in mode-locking mode with a picosecond light pulse duration. This extends the results of the research to other laser projection systems that use acousto-optical modulators on paratellurite.

The study of the characteristics of acousto-optical systems with a pulsed method of forming a line for displaying and recording information using copper vapor lasers makes it possible to extend the results of this study to systems using full-colour pulsed solid-state lasers, which are currently developing rapidly.

Appendix 1

Parameters of some crystals used in acousto-optical devices

Parameters of lithium niobate crystals LiNbO$_3$

Lithium niobate is a colourless ferroelectric single crystal grown by the Czochralski method. Generally, crystals of congruent composition are grown when the compositions of the melt and the crystal grown from it are the same: 48.6 mol% Li$_2$O and 51.4 mol% Nb$_2$O$_5$. According to the equilibrium diagram, there is also a stoichiometric composition of this compound, whose crystals with a composition of 50:50 mol% of the starting oxides can be grown from a melt containing 58 mol% Li$_2$O and 42 mol% Nb$_2$O$_5$, respectively. Such a large difference in the composition of the melt and crystal will lead to significant variations in the crystal composition throughout the entire growing process, which affects the reproducibility of the physical parameters.

The lithium niobate of stoichiometric composition has a more ordered structure in comparison with the congruent LiNbO$_3$, which leads to a change in some of its properties, for example, the photorefractive stability increases by two or three orders of magnitude, the induced IR absorption and also the coercive field are significantly reduced. The degree of perfection of the structure is considered to be the content of Li$_2$O.

To reduce the 'optical damage' effect, the crystals of the congruent composition are doped with magnesium. When introducing into the melt up to 6 mol% Mg, there is practically no optical damage in the grown crystals.

For lithium niobate piezoelectric converters, congruent crystals are usually used for acousto-optical devices. The main requirements for such crystals are the single domain, the absence of bubbles and mechanical stresses. Below are the main parameters of lithium niobate crystals of congruent composition [87, 209, 210].

Physical properties of LiNbO₃ crystal

Symmetry trigonal, 3m (R_{3c}).
The lattice parameters $a = 5.148$ Å; $c = 13.863$ Å.
Molecular weight: 147.9.
Density: 4,644 g/cm³.
Melting point: 1530 K.
Curie temperature: 1415 K.
Moss hardness: 5.
Coefficients of thermal expansion at 293 K: $\alpha_{11} = 15 \times 10^{-6}$ K⁻¹; $\alpha_{22} = 15 \cdot 10^{-6}$ K⁻¹; $\alpha_{33} = 5 \cdot 10^{-6}$ K⁻¹.
The heat capacity 0.15 cal/(g K).
Thermal conductivity 56 mW/cm · K.
Electro-optical coefficients for He–Ne laser radiation 632.8 nm:
$r_{33} = 30.9 \cdot 10^{-12}$ m/V, $r_{13} = r_{23} = 9.6 \cdot 10^{-12}$ m/V,
$r_{22} = -r_{12} = -r_{61} = 6.8 \cdot 10^{-12}$ m/V, $r_{51} = r_{42} = 32.6 \cdot 10{-12}$ m/V.
Dielectric constant: $\varepsilon^T_{11} = 82.5$; $\varepsilon^T_{33} = 28.3$.
Piezoelectric modules:
$d_{22} = 20.95 \cdot 10^{-12}$ C/H, $d_{15} = 65.36 \cdot 10^{-12}$ C/H,

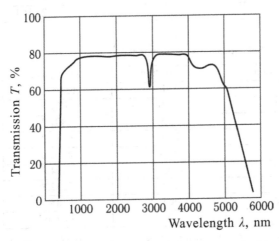

Fig. A.1.1. The dependence of the transmission of optical radiation $T,\%$ on the wavelength λ for a LiNbO₃ crystal without an antireflection coating.

Table A.1.1. The refractive indices of LiNbO$_3$

Radiation wavelength, λ, µm	Refractive indices	
	Ordinary n_o	Unusual n_z
0.42	2.4144	2.3038
0.45	2.3814	2.2765
0.5	2.3444	2.2446
0.55	2.3188	2.2241
0.6	2.3002	2.2083
0.65	2.2862	2.1964
0.7	2.2756	2.19
0.8	2.2598	2.1741
0.9	2.2487	2.1647
1	2.2407	2.158
1.2	2.2291	2.1481
1.4	2.2208	2.141
1.6	2.2139	2.1351
1.8	2.2074	2.1279
2	2.2015	2.1244
2.2	2.1948	2.1187
2.4	2.1882	2.1138
2.6	2.1814	2.108
2.8	2.1741	2.102
3	2.1663	2.0955
3.2	2.158	2.0886
3.4	2.1493	2.0814
3.6	2.1398	2.0735
3.8	2.1299	2.0652
4	2.1193	2.0564

Table A.1.2. Velocities of elastic waves *

Direction	Polarization	Speed, m/s
X	X, L	6548.73
X	Y, S	4059.3
X	Z, S	4801.2
Y	Y, L	6883.2
Y	X, S	3961.52
Y	Z, QS	4494.3
Z	Z, L	7332.8
Z	X or Y ,S	3588.5
Y	S	4466,67

* Types of elastic waves: L – longitudinal; S – transverse; QS – quasitransverse.

Table A.1.3. Acousto-optical properties for the most used geometries of acousto-optical interaction ($\lambda = 0.6328$ μm)

The direction of propagation of elastic waves	Polarization of elastic waves	Direction of light	Polarization of light	Parameters of acousto-optical quality		
				$M_1 \cdot 10^{-8}$ s³/kg	$M_2 \cdot 10^{-15}$ s³/kg	$M_3 \cdot 10^{-11}$ s³/kg
				$n^7 p^2/\rho V$	$n^6 p^2/\rho V^3$	$n^7 p^2/\rho V^2$
Z	Z	X	Y	10.2	0.08	1.4
Z	Z	$\perp Z$	Z	4.46	0.38	0.6
X	X	Y	Z	24.3	2.6	3.7
X	X	$\sim Y + 41°$	$\sim Z - 41°$	66.9*	6.9*	10.2*
X	X	Z	Y	6.6	0.67	1
Z	Y	X	Y	44.4	15.1	12.4
X	X	$\perp X$	X	0.95	0.1	0.15

* Maximum efficiency for diffraction of light on longitudinal waves in the X-direction.

Table A.1.4. Parameters of piezoelectric converters from LiNbO$_3$

Crystal	Cut	Mode	Coefficient of electromechanical coupling	Density, kg/m³	Speed, V, m/s	Impedance $Z_0 = \rho V$ 10^6, kg/(m²s)	Resonance frequency, $V/2$, MHz/μm	Relative permittivity
LiN-bO$_3$ (3m)	36° Y	L	0,49	4640	7300	33,9	3650	38,6
	X	S	0,68		4800	22,3	2400	44,3
	163° Y	S	0,62		4480	20,8	2240	42,9
	Z	L	0,17		7320	34,0	3660	29

$d_{31} = -1.32 \cdot 10^{-12}$ C/H, $d_{33} = 8.27 \cdot 10^{-12}$ C/H.

Elastic constants, $\cdot 10^{-10}$ N/m²: $c_{11} = c_{22} = 20.18$; $c_{12} = c_{21} = 5.56$; $c_{13} = c_{23} = c_{31} = c_{32} = 7.49$; $c_{14} = c_{41} = c_{56} = c_{65} = 0.83$; $c_{24} = c_{42} = -0.83$; $c_{33} = 24.13$; $c_{44} = c_{55} = 5.95$; c66 = 7.28.

Photoelastic constants for the emission of helium from a neon laser $\lambda = 0.6328$ μm:

$p_{11} = p_{22} = 0.03$; $p_{12} = p_{21} = p_{44} = 0.079$;

$p_{13} = p_{23} = 0.104$; $p_{31} = p_{32} = 0.172$;

$p_{14} = p_{65} = -0.071$; $p_{24} = 0.071$; $p_{41} = p_{56} = -0.152$; $p_{42} = 0.152$;

$p_{33} = 0.078$; $p_{44} = p_{55} = 0.22$; $p_{66} = -0.025$.

Optical transparency in the range of 0.4–5.5 μm.

The parameters of the sodium molybdate crystal of NaBi $(MoO_4)_2$

Molybdate sodium bismuth is a yellowish single crystal grown by the Czochralski method. This is a new crystal grown for the first time, replacing the crystal of lead molybdate. The method of cultivation is protected by copyright certificates [117, 118]. The advantage of the NaBi $(MoO_4)_2$ bismuth sodium molybdate crystal is high

Table A.1.5. Parameters of the piezotransducer of the 36°Y-cut LiNbO$_3$

Central frequency f_0, MHz	Thickness $t = 0.5V/f_0$, µm	Capacity C, pF/mm^2	Specific resistance $1/(\omega_0 C)$, Ohm/mm^2	The electrode area Δ for $(\omega_0 C = 50\ \Omega)$, mm^2
30	121.7	2.81	1887	37.8
40	91.3	3.74	1064	21.3
50	73.0	4.68	681	13.6
60	60.8	5.62	472	9.44
70	52.1	6.56	347	6.94
80	45.6	7.49	266	5.32
90	40.6	8.41	210	4.20
100	36.5	9.36	170	3.40
110	33.2	10.3	141	2.82
120	30.4	11.2	118	2.36
130	28.1	12.2	100	2.00
140	26.1	13.1	86.8	1.74
150	24.3	14.1	75.5	1.51
160	22.8	15.0	66.3	1.33
180	20.3	16.8	52.7	1.05
200	18.3	18.7	42.6	0.852
250	14.6	23.4	27.2	0.544
300	12.2	28.1	189	0.380
350	10.4	32.9	13.8	0.276
400	9.13	37.4	10.6	0.212
450	8.11	42.1	8.41	0.168
500	7.30	46.8	6.81	0.136
1000	3.65	93.6	1.70	$3.40 \cdot 10^{-2}$
1500	2.43	141	0.755	$1.51 \cdot 10^{-2}$
2000	1.83	187	0.426	$8.52 \cdot 10^{-3}$
2500	1.46	234	0.272	$5.44 \cdot 10^{-3}$
3000	1,22	280	0,190	$3.80 \cdot 10^{-3}$

Table A.1.6. Parameters of the X-cut piezoelectric transducer LiNbO$_3$

Central frequency, f_0, MHz	Thickness $t = 0{,}5V/f_0$, µm	Capacity C, pF/mm^2	Specific resistance $1/(\omega_0 C)$, Ohm/mm^2	The electrode area Δ for $(\omega_0 C = 50\ \Omega)$, mm^2
30	80.0	4.90~	1083	21.7
40	60.0	6.53	610	12.2
50	48.0	8.17	390	7.80
60	40.0	9.80	271	5.42
70	34.3	11.4	200	4.00
80	30.0	13.1	152	3.04
90	26.7	14.7	120	2.40
100	24.0	16.3	97.7	1.95
110	21.8	18.0	80.4	1.61
120	20.0	19.6	67.7	1.35
130	18.5	21.3	57.8	1.16
140	17.1	22.9	49.7	0.94
150	16.0	24.5	43.3	0.866
160	15.0	26.1	38.1	0.762
180	13.3	29.5	30.0	0.600
200	12.0	32.7	24.4	0.488
250	9.60	40.8	15.6	0.312
300	8.00	49.0	10.8	0.217
350	6.86	57.2	795	0.159
400	6.00	65.3	6.10	0.122
450	5.33	73.5	4.82	0.0964
500	4.80	81.7	3.90	0.078
1000	2.40	163	0.977	$1.95 \cdot 10^{-2}$
1500	1.60	245	0.433	$8.66 \cdot 10^{-3}$
2000	1.20	327	0.244	$4.88 \cdot 10^{-3}$
2500	0.96	408	0.156	$3.12 \cdot 10^{-3}$
3000	0,80	490	0,108	$2{,}17 \cdot 10^{-3}$

processability in the manufacture of acousto-optical devices, as well as its radiation resistance, optical quality and rather large sizes of grown bulbs [117, 118, 211–214].

The physical properties of the NaBi (MoO$_4$)$_2$ crystal

Symmetry: tetragonal, 422 (4/m) – structure of scheelite type.
    Lattice parameters: $a = 5.267$ Å; $c = 11.552$ Å.

Molecular weight: 551.845.

Density: 5.71 ± 0.02 g/cm³.

Melting point: 840°C.

Moss hardness: 4–5.

Coefficients of thermal expansion at 293 K:

$\alpha_{11} = 9.7 \cdot 10^{-6}$ K^{-1}; $\alpha_{33} = 21.7 \cdot 10^{-6}$ K^{-1}.

Thermal conductivity: 30 mW/cm · K.

Optical transparency range: 0.45–6.0 μm.

The induced electrohydration coefficient at a wavelength of $\lambda =$ 0.6328 μm with an electric field strength $E = 10$ kV/cm and room temperature $T = 20°C$: 0.12 deg/mm.

Maximum permissible electric field strength: $E = 100$–120 kV/ cm).

Dielectric permeability: $\varepsilon_{11} = 37$; $\varepsilon_{33} = 42$.

Elastic constants, $\cdot 10^{-10}$ N/m²: $c_{11} = 10.674$;

$c_{33} = 8.84$; $c_{44} = 2.57$; $c_{66} = 3.576$;

$c_{12} = 5.197$; $c_{13} = 3.565$; $c_{16} = -1.133$.

Photoelastic constants:

$P_{11} = P_{22} = 0.195$; $P_{12} = P_{21} = 0.191$; $P_{13} = P_{23} = 0.184$;

$P_{31} = P_{32} = 0.178$;

$P_{33} = 0.184$; $P_{44} = P_{55} = 0.023$; $P_{45} = -0.010$;

$P_{66} = 0.019$; $P_{61} = -0.018$; $P_{16} = 0.021$.

Attenuation of longitudinal acoustic waves with a frequency of 500 MHz propagating along the optical axis is –0.6 dB/μs.

Table A.1.7. The refractive indices of NaBi (MoO$_4$)$_2$

Radiation wave-length λ, μm	Refractive indices		$\Delta n = n_e - n_o$
	Ordinary n_o	Unusual n_e	
0.4046	2.509	2.341	−0.168
0.4861	2.417	2.285	−0.132
0.4890	2.408	2.278	−0.130
0.4922	2.396	2.270	−0.126
0.5461	2.353	2.241	−0.112
0.5770	2.332	2.227	−0.105
0.5791	2.331	2.226	−0.105
0.5876	2.327	2.223	−0.104
0.5892	2.326	2.223	−0.103
0.6438	2.307	2.209	−0.098
0.6562	2.297	2.202	−0.095
0.7065	2.290	2.196	−0.094

Table A.1.8. Velocity of elastic waves in NaBi (MoO$_4$)$_2$

Direction of propagation of elastic waves	Polarization of elastic waves	Velocity of elastic waves · 10^3 m/s
[100]	[100]	4.360
[100]	[010]	2.440
[100]	[001]	2.100
[001]	[100]	2.110
[001]	[010]	2.110
[001]	[001]	3.935
[101]	[101]	4.040
[101]	[101]	2.450
[101]	[010]	2.131
[110]	[110]	4.546
[110]	[110]	2.081
[110]	[001]	2,154

Table A.1.9. Acousto-optical properties of NaBi (MoO$_4$)$_2$

Direction of propagation of longitudinal elastic waves	Direction of light propagation	Polarization of light	Acousto-optical quality M_2, · 10^{-15} s^3/kg
[001]	[100]	[001]	23.72
[001]	[100]	[010]	2.61
[100]	[010]	[001]	0.11
[100]	[001]	[010]	2.51
[100]	[010]	[100]	7.6

Parameters of crystals of paratellurite TeO$_2$

Paratellurite is a colourless single crystal grown by the Czochralski method using the echnology (Author Cert. No. 1529785 [114]), has high radiation strength, high optical quality, and the absence of polarization.

The physical properties of the TeO$_2$ crystal [130, 215-217]

Symmetry: tetragonal, 422 (D4)
Lattice parameters: $a = 4.8122$ Å; $c = 7.6157$ Å.
Molecular weight: 159.51.

Table A.1.10. Refractive indices TeO$_2$

Wavelength	Refraction parameters		The refractive index difference
λ, μm	Ordinary	Unusual	
	n_o	n_e	$\Delta n = n_e - n_o$
0.4047	2.4315	2.6167	0.1852
0.4358	2.3834	2.5583	0.1749
0.4678	2.3478	2.5164	0.1686
0.4800	2.3366	2.5036	0.1670
0.5086	2.3150	2.4779	0.1629
0.5461	2.2931	2.4520	0.1589
0.5893	2.2738	2.4295	0.1557
0.6328	2.2597	2.4119	0.1522
0.6438	2.2562	2.4086	0.1524
0.690	2.2450	2.3955	0.1505
0.800	2.226	2.373	0.147
1.00	2.208	2.352	0.144
1.31	2.19	2.33	0.14
1.55	2.179	2.317	0.138
2	2.18	2.316	0.136
2.4	2.155	2.289	0.136
3		2.293	
3.5	2.15	2.281	0.131
4	2.121	2.252	0.131
5	2.087	2.213	0.126
6	2.042	2.159	0.117

Density: 5.99 ± 0.03 g/cm^3.
Melting point: 733°C.
Moss hardness: 3–4.
Coefficients of thermal expansion at 293 K:
$\alpha_{11} = 17.7 \cdot 10^{-6}$ K^{-1}; $\alpha_{22} = 17.7 \cdot 10^{-6}$ K^{-1}; $\alpha_{33} = 5.5 \cdot 10^{-6}$ K^{-1}.
Thermal conductivity, 30 mW/cm K.
Dielectric permeability: $\varepsilon_{11} = 22.9$; $\varepsilon_{33} = 24.7$.
Elastic constants, $\cdot 10^{-10}$ N/m^2: $c_{11} = 5.57$; $c_{33} = 10.58$; $c_{44} = 2.65$; $c_{66} = 6.59$; $c_{12} = 5.12$; $c_{13} = 2.18$.
Photoelastic constants ($\lambda = 0.6328$ μm): $p_{11} = 0.0074$; $p_{12} = 0.187$; $p_{13} = 0.340$; $p_{31} = 0.0905$; $p_{33} = 0.240$; $p_{44} = -0.17$; $p_{66} = -0.0463$.
Optical transparency range: 0.35–5.0 μm.

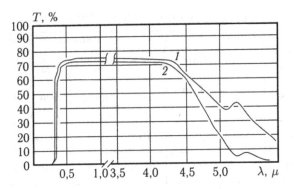

Fig. A.1.2. Dependence of the transmission T, % on the optical wavelength λ for the paratellurite TeO_2 crystal, which does not have antireflection coatings: *1* – for radiation polarized perpendicular to the optical axis; curve *2* – for radiation polarized parallel to the optical axis. Sample thickness 20 mm.

Table A.1.11. The optical activity along the [001] direction in TeO_2

Wavelength λ, μm	Specific rotation ρ, deg/mm
0.3698	587.1
0.3783	520.6
0.3917	437.4
0.4152	337.6
0.4382	271.0
0.4630	221.1
0.4995	171.2
0.5300	143.4
0.5893	104.9
0.6328	86.9
0.700	67.4
0.800	48.5
0.900	37.4
1.0	29.5
1.1	23.8
1.2	22.5
1.31	18.5
1.5	15.4
1.55	14.5
2	10.5
2.4	8.4

Table A.1.12. Acousto-optical properties of TeO$_2$ ($\lambda = 0.6328$ μm)

Direction of diffusion elastic waves	Polarization of elastic waves	Speed of elastic waves, $\cdot 10^3$ m/s	Direction of light	Polarization of light	Parameters of acousto-optical quality		
					M_1, $\cdot 10^{-8}$ m^{2}s/kg $n^7p^2/\rho V$	M_2, $\cdot 10^{-15}$ s^3/kg $n^6p^2/\rho V^3$	M_3, $\cdot 10^{-11}$ ms^2/kg $n^7p^2/\rho V^2$
[100]	[100]	2.98	[010]	[100]	0.097	0.048	0.031
[100]	[100]	2.98	[010]	[001]	22.9	10.6	7.20
[001]	[001]	4.26	[010]	[100]	149	34.5	33.1
[001]	[001]	4.26	[010]	[001]	113	25.6	26.4
[100]	[010]	3.04	[001]	arbitrary	3.7	1.76	1.2
[110]	[110]	4.21	[$\bar{1}$10]	[110]	6.54	1.6	1.5
[110]	[110]	4.21	[$\bar{1}$10]	[001]	16.2	3.77	3.8
[101]	[101]	3.64	[$\bar{1}$01]	[010]	101	33.4	27.5
[010]	[010]	2.98	[$\bar{1}$01]	[101]	42.6	20.4	14.6
[110]	[$\bar{1}$10]	0.617	[001]	arbitrary	68.6	793	110
[101]	[$\bar{1}$01]	2.08	[010]	[100]	76.4	77	36
[110]	[$\bar{1}$10]	0.617	[001]	circular	103	1200	168

Appendix 2

Expressions for the time-average light intensity in the fifth and fourth order of interaction

In the approximation of the fourth and fifth orders of interaction, the calculation of the field is made according to the scattering diagram shown in Fig. 2.22.

The time-average intensity of light in the image signal for modulation of ultrasound by the harmonic law in the 0-th diffraction order and in the fourth-order approximation of the interaction has the form:

$$
I_0^{(4)}(x_1) = \tilde{\Gamma}_0 \cdot \left\{ C_0 + 2 \cdot \sum_{n=1}^{8} \left[C_{n1} \cdot \cos\left(\frac{2\pi \cdot n \cdot f_0 \cdot x_1}{v}\right) + \right. \right.
$$
$$
\left. \left. + C_{n2} \cdot \cos\left(\frac{2\pi \cdot n \cdot f_0 \cdot x_1}{v}\right) \cdot e^{-\left(\frac{\pi \cdot n \cdot f_0 \cdot \tau_0}{2\sqrt{\ln 2}}\right)^2} \right] \right\},
$$

(A.2.1)

where

$$
C_0 = \sum_{k=1}^{2} \sum_{l=-4}^{4} (A_{lk}^2 + B_{lk}^2),
$$

(A.2.2)

$$
C_{n1} = \sum_{k=1}^{2} \sum_{l=-4}^{4} \sum_{m=-4}^{4} (A_{lk} \cdot A_{mk} + B_{lk} \cdot B_{mk}) \cdot \delta(l-m-n),
$$

(A.2.3)

$$
C_{n2} = \sum_{k=1}^{2} \sum_{l=-4}^{4} \sum_{m=-4}^{4} (A_{lk} \cdot B_{mk} - B_{lk} \cdot A_{mk}) \cdot \delta(l-m-n),
$$

(A.2.4)

$$A_{l1} = (-1)\cdot(\sin\varphi(\theta_l)\cdot D_l(\theta_l) - \rho(\theta_l)\cdot\cos\varphi(\theta_l)\cdot H_l(\theta_l)),$$
$$B_{l1} = (-1)\cdot(\rho(\theta_l)\cdot\cos\varphi(\theta_l)\cdot D_l(\theta_l) + \sin\varphi(\theta_l)\cdot H_l(\theta_l)),$$
$$A_{l2} = \cos\varphi(\theta_l)\cdot D_l(\theta_l) + \rho(\theta_l)\cdot\sin\varphi(\theta_l)\cdot H_l(\theta_l), \tag{A.2.5}$$
$$B_{l2} = \cos\varphi(\theta_l)\cdot H_l(\theta_l) - \rho(\theta_l)\cdot\sin\varphi(\theta_l)\cdot D_l(\theta_l),$$
$$D_l = \text{Re}(\tilde{S}_l), \quad H_l = \text{Im}(\tilde{S}_l),$$

$$\tilde{S}_0 = 1 - \frac{\chi_0^2}{2}\left[S(0,0) + \frac{m_0^2}{4}\cdot S(0,\pm1)\right] + \frac{\chi_0^4}{24}\left\{ S(0,0,0,0) + \frac{m_0^2}{4}\times\right.$$
$$\times[S(0,\pm1,\pm1,\pm1) + S(0,\pm1,\pm1,0) + S(0,\pm1,0,0) + S(0,0,\pm1,\pm1) +$$
$$+ S(0,0,0,\pm1) + S(0,0,\pm1,0)] + \frac{m_0^4}{16}[S(0,\pm1,\pm2,\pm1) + \tag{A.2.6}$$
$$\left.+ S(0,\pm1,0,\pm1) + S(0,\pm1,0,\mp1)]\right\},$$

$$\tilde{S}_{\pm1} = -\frac{\chi_0^2}{2}\cdot\frac{m_0}{2}\cdot[S(\pm1,0) + S(\pm1,\pm1)] + \frac{\chi_0^4}{24}\cdot\frac{m_0^2}{4}\cdot\{S(\pm1,\pm1,\pm1,\pm1) +$$
$$+ S(\pm1,\pm1,\pm1,0) + S(\pm1,\pm1,0,0) + S(\pm1,0,0,0) +$$
$$+ \frac{m_0^2}{4}\cdot[S(\pm1,\pm2,\pm2,\pm1) + S(\pm1,\pm2,\pm1,\pm1)] + S(\pm1,\pm2,\pm1,0) + \tag{A.2.7}$$
$$+ S(\pm1,\pm1,\pm2,\pm1) + S(\pm1,\pm1,0,\pm1) + S(\pm1,\pm1,0,\mp1) +$$
$$+ S(\pm1,0,\pm1,\pm1) + S(\pm1,0,0,\pm1) + S(\pm1,0,\mp1,0) +$$
$$+ S(\pm1,0,\pm1,0) + S(\pm1,0,\mp1,\mp1) + S(\pm1,0,0,\mp1)]\},$$

$$\tilde{S}_{\pm2} = -\frac{\chi_0^2}{2}\cdot\frac{m_0^2}{4}\cdot S(\pm2,\pm1) + \frac{\chi_0^4}{24}\cdot\frac{m_0^2}{4}\cdot\{S(\pm2,\pm2,\pm2,\pm1) +$$
$$+ S(\pm2,\pm2,\pm1,\pm1) + S(\pm2,\pm2,\pm1,0) + S(\pm2,\pm1,\pm1,\pm1) +$$
$$+ S(\pm2,\pm1,0,0) + \frac{m_0^2}{4}\cdot[S(\pm2,\pm3,\pm2,\pm1) + S(\pm2,\pm1,\pm2,\pm1) + \tag{A.2.8}$$
$$+ S(\pm2,\pm1,0,\pm1) + S(\pm2,\pm1,0,\mp1)]\},$$

$$\tilde{S}_{\pm3} = \frac{\chi_0^2}{24}\cdot\frac{m_0^3}{8}\cdot[S(\pm3,\pm3,\pm2,\pm1) + S(\pm3,\pm2,\pm2,\pm1) +$$
$$+ S(\pm3,\pm2,\pm1,\pm1) + S(\pm3,\pm2,\pm1,0), \tag{A.2.9}$$

$$\tilde{S}_{\pm4} = \frac{\chi_0^2}{24}\cdot\frac{m_0^3}{16}\cdot S(\pm4,\pm3,\pm2,\pm1), \tag{A.2.10}$$

$$S(m,n) = \Upsilon(\beta_{mn}^{(2)}) \cdot \Upsilon(\beta_{n0}^{(1)}) \cdot K_{21}(\theta_m^{(2)}, \theta_n^{(1)}) \cdot K_{12}(\theta_n^{(1)}, \theta_0^{(0)}), \qquad (A.2.11)$$

$$S(p,m,n) = \Upsilon(\beta_{mn}^{(3)}) \cdot K_{12}(\theta_p^{(3)}, \theta_m^{(2)}) \cdot S(m,n), \qquad (A.2.12)$$

$$S(l,p,m,n) = \Upsilon(\beta_{lp}^{(4)}) \cdot K_{21}(\theta_l^{(4)}, \theta_p^{(3)}) \cdot S(p,m,n), \qquad (A.2.13)$$

$$\tilde{\Gamma}_0 = \frac{\tilde{n} \cdot \tau_0 \cdot a_0^2 \cdot E_0^2}{16 \cdot \lambda_0 \cdot \sqrt{\pi \cdot \ln 2} \cdot M_1' \cdot M_2' \cdot T \cdot F_1}. \qquad (A.2.14)$$

The time-average intensity of light in the image signal when ultrasound is modulated according to the harmonic law in the + 1-m diffraction order and in the approximation of the fifth order of interaction has the form:

$$I_{+1}^{(5)}(x_1) = \tilde{\Gamma}_0 \left\{ \tilde{C}_0 + 2 \cdot \sum_{n=1}^{10} [\tilde{C}_{n1} \cdot \cos\left(\frac{2\pi \cdot n \cdot f_0 \cdot x_1}{v}\right) + \right.$$

$$\left. + \tilde{C}_{n2} \cdot \sin\left(\frac{2\pi \cdot n \cdot f_0 \cdot x_1}{v}\right) \cdot e^{-\left(\frac{\pi \cdot n \cdot f_0 \cdot \tau_0}{2\sqrt{\ln 2}}\right)^2} \right\}, \qquad (A.2.15)$$

where

$$\tilde{C}_0 = \sum_{k=1}^{2} \cdot \sum_{l=-5}^{5} (A_{lk}^2 + B_{lk}^2), \qquad (A.2.16)$$

$$\tilde{C}_{n1} = \sum_{k=1}^{2} \cdot \sum_{l=-5}^{5} \sum_{m=-5}^{5} (A_{lk} \cdot A_{mk} + B_{lk} \cdot B_{mk}) \cdot \delta(l-m-n), \qquad (A.2.17)$$

$$\tilde{C}_{n2} = \sum_{k=1}^{2} \cdot \sum_{l=-5}^{5} \sum_{m=-5}^{5} (A_{lk} \cdot B_{mk} + B_{lk} \cdot A_{mk}) \cdot \delta(l-m-n), \qquad (A.2.18)$$

$$\begin{aligned}
A_{l1} &= \cos\varphi(\theta_l) \cdot D_l(\theta_l) - \rho(\theta_l) \cdot \sin\varphi(\theta_l) \cdot H_l(\theta_l), \\
B_{l1} &= \cos\varphi(\theta_l) \cdot H_l(\theta_l) + \rho(\theta_l) \cdot \sin\varphi(\theta_l) \cdot D_l(\theta_l), \\
A_{l2} &= \sin\varphi(\theta_l) \cdot D_l(\theta_l) + \rho(\theta_l) \cdot \cos\varphi(\theta_l) \cdot H_l(\theta_l), \qquad (A.2.19) \\
B_{l2} &= \sin\varphi(\theta_l) \cdot H_l(\theta_l) - \rho(\theta_l) \cdot \cos\varphi(\theta_l) \cdot D_l(\theta_l), \\
D_l &= \mathrm{Re}(\tilde{S}_l), \quad H_l = \mathrm{Im}(\tilde{S}_l),
\end{aligned}$$

Appendix 2

$$S(q,l,p,m,n) = \Upsilon(\beta_{ql}^{(5)}) \cdot K_{12}(\theta_q^{(5)}, \theta_l^{(4)}) \cdot S(l,p,m,n), \qquad (A.2.20)$$

$$\tilde{S}_0 = K_{12}(\theta_0^{(1)}, \theta_0^{(0)}) - \frac{\chi_0^2}{6} \left\{ S(0,0,0) + \frac{m_0^2}{4} \cdot \right.$$

$$[S(0,0,\pm 1) + S(0,\pm 1,\pm 1) +$$

$$S(0,\pm 1,0)]\} + \frac{\chi_0^4}{120} \left[\!\!\left[S(0,0,0,0,0) + \frac{m_0^2}{4} \cdot \{S(0,\pm 1,\pm 1,\pm 1,\pm 1) + \right.\right.$$

$$S(0,\pm 1,\pm 1,\pm 1,0) + S(0,\pm 1,\pm 1,0,0) + S(0,\pm 1,0,0,0) +$$

$$S(0,0,\pm 1,\pm 1,\pm 1) + S(0,0,\pm 1,\pm 1,0) + S(0,0,\pm 1,0,0) +$$

$$S(0,0,0,\pm 1,\pm 1) + S(0,0,0,\pm 1,0) + S(0,0,0,0,\pm 1) +$$

$$\frac{m_0^4}{4} \cdot [S(0,\pm 1,\pm 2,\pm 2,\pm 1) + S(0,\pm 1,\pm 2,\pm 1,\pm 1) +$$

$$S(0,\pm 1,\pm 2,\pm 1,0) + S(0,\pm 1,\pm 1,\pm 2,\pm 1) + S(0,\pm 1,\pm 1,0,\pm 1) +$$

$$S(0,\pm 1,\pm 1,0,\mp 1) + S(0,\pm 1,0,\pm 1,\pm 1) + S(0,\pm 1,0,\pm 1,0) +$$

$$S(0,\pm 1,0,0,\pm 1) + S(0,\pm 1,0,0,\mp 1) + S(0,\pm 1,0,\mp 1,0) +$$

$$S(0,\pm 1,0,\mp 1,\mp 1) + S(0,0,\pm 1,\pm 2,\pm 1) + S(0,0,\pm 1,0\pm 1) +$$

$$S(0,0,\pm 1,0,\mp 1)]\} \left.\left.\right]\!\!\right]$$

(A.2.21)

$$\tilde{S}_{\pm 2} = \frac{m_0^2}{4} \cdot \left[\!\!\left[-\frac{\chi_0^2}{6} \cdot [S(\pm 2,\pm 2,\pm 1) + S(\pm 2,\pm 1,\pm 1) + S(\pm 2,\pm 1,0)] + \right.\right.$$

$$+\frac{\chi_0^4}{120} \cdot \{S(\pm 2,\pm 2,\pm 2,\pm 2,\pm 1) + S(\pm 2,\pm 2,\pm 1,\pm 1,\pm 1) +$$

$$+S(\pm 2,\pm 2,\pm 1,\pm 1,0) + S(\pm 2,\pm 2,\pm 1,0,0) + S(\pm 2,\pm 1,\pm 1,\pm 1,\pm 1) +$$

$$+S(\pm 2,\pm 1,\pm 1,\pm 1,0) + S(\pm 2,\pm 1,0,0,0) +$$

$$+\frac{m_0^2}{4} \cdot [S(\pm 2,\pm 3,\pm 3,\pm 2,\pm 1) + S(\pm 2,\pm 3,\pm 2,\pm 2,\pm 1) +$$

(A.2.22)

$$+S(\pm 2,\pm 3,\pm 2,\pm 1,0) + S(\pm 2,\pm 3,\pm 2,\pm 1,\pm 1) + S(\pm 2,\pm 2,\pm 3,\pm 2,\pm 1) +$$

$$+S(\pm 2,\pm 2,\pm 1,\pm 2,\pm 1) + S(\pm 2,\pm 2,\pm 1,0,\pm 1) + S(\pm 2,\pm 2,\pm 1,0\pm 1) +$$

$$+S(\pm 2,\pm 2,\pm 1,0,\mp 1) + S(\pm 2,\pm 1,\pm 2,\pm 2,\pm 1) + S(\pm 2,\pm 1,\pm 1,\pm 2,\pm 1) +$$

$$+S(\pm 2,\pm 1,0,\pm 1,\pm 1) + S(\pm 2,\pm 1,0,\pm 1,0) + S(\pm 2,\pm 1,0,0,\pm 1) +$$

$$+S(\pm 2,\pm 1,0,0,\mp 1) + S(\pm 2,\pm 1,0,\mp 1,0) + S(\pm 2,\pm 1,0,\mp 1,\mp 1)]\} \left.\left.\right]\!\!\right]$$

$$\tilde{S}_{\pm 4} = \frac{m_0^2}{16} \frac{\chi_0^4}{120} \cdot [S(\pm 4, \pm 4, \pm 3, \pm 2, \pm 1) + S(\pm 4, \pm 3, \pm 3, \pm 2, \pm 1) +$$

$$+ S(\pm 4, \pm 3, \pm 2, \pm 2, \pm 1) + S(\pm 4, \pm 3, \pm 2, \pm 1, \pm 1) + S(\pm 4, \pm 3, \pm 2, \pm 1, 0)] \tag{A.2.23}$$

$$\tilde{S}_{\pm 1} = \frac{m_0^2}{2} \left\{ K_{12}(\theta_{\pm 1}^{(1)}, \theta_0^{(0)}) \cdot \Upsilon(\beta_{\pm 1,0}^{(1)} - \frac{\chi_0^2}{6} \cdot \{S(\pm 1, \pm 1, \pm 1) + \right.$$

$$+ S(\pm 1, \pm 1, 0) + S(\pm 1, 0, 0) + \frac{m_0^2}{4} \cdot [S(\pm 1, 0, \pm 1) + S(\pm 1, 0, \mp 1) +$$

$$+ S(\pm 1, \pm 2, \pm 1)]\} + \frac{\chi_0^4}{4} \cdot \left[\! \left[S(\pm 1, \pm 1, \pm 1, \pm 1, \pm 1) + S(\pm 1, \pm 1, \pm 1, \pm 1, 0) + \right.\right.$$

$$+ S(\pm 1, \pm 1, 0, 0, 0) + S(\pm 1, 0, 0, 0, 0) + \frac{m_0^2}{4} \cdot \{S(\pm 1, \pm 2, \pm 2, \pm 2, \pm 1) +$$

$$+ S(\pm 1, \pm 2, \pm 2, \pm 1, \pm 1) + S(\pm 1, \pm 2, \pm 2, \pm 1, 0) + S(\pm 1, \pm 2, \pm 1, \pm 1, \pm 1) +$$

$$+ S(\pm 1, \pm 2, \pm 1, \pm 1, 0) + S(\pm 1, \pm 2, \pm 1, 0, 0) + S(\pm 1, \pm 1, \pm 2, \pm 2, \pm 1) +$$

$$+ S(\pm 1, \pm 1, \pm 2, \pm 1, \pm 1) + S(\pm 1, \pm 1, \pm 2, \pm 1, 0) + S(\pm 1, \pm 1, \pm 1, \pm 2, \pm 1) + \tag{A.2.24}$$

$$+ S(\pm 1, \pm 1, 0, \pm 1, \pm 1) + S(\pm 1, \pm 1, 0, \pm 1, 0) + S(\pm 1, \pm 1, 0, 0, \pm 1) +$$

$$+ S(\pm 1, \pm 1, 0, 0, \mp 1) + S(\pm 1, 0, \pm 1, \mp 1, 0) + S(\pm 1, \pm 1, 0, \mp 1, \mp 1) +$$

$$+ S(\pm 1, 0, \pm 1, \pm 1, \pm 1) + S(\pm 1, 0, \pm 1, \pm 1, 0) + S(\pm 1, 0, \pm 1, 0, 0) +$$

$$+ S(\pm 1, 0, 0, \pm 1, \pm 1) + S(\pm 1, 0, 0, \pm 1, 0) + S(\pm 1, 0, 0, 0, \pm 1) +$$

$$+ S(\pm 1, 0, 0, 0, \mp 1) + S(\pm 1, 0, 0, \pm 1, 0) + S(\pm 1, 0, 0, \mp 1, \mp 1) +$$

$$+ \frac{m_0^2}{4} \cdot [S(\pm 1, \pm 2, \pm 3, \pm 2, \pm 1) + S(\pm 1, \pm 2, \pm 1, \pm 2, \pm 1) +$$

$$+ S(\pm 1, \pm 2, \pm 1, 0, \pm 1) + S(\pm 1, \pm 2, \pm 1, 0, \mp 1) + S(\pm 1, 0, \pm 1, \pm 2, \pm 1) +$$

$$\left.\left.+ S(\pm 1, 0, \pm 1, 0, \pm 1) + S(\pm 1, 0, \pm 1, 0, \mp 1)] \right\} \right]\! \right]\Big\}$$

$$\tilde{S}_{\pm 3} = \frac{m_0^2}{8} \cdot \left[\! \left[-\frac{\chi_0^2}{6} \cdot S(\pm 3, \pm 2, \pm 1) + \frac{\chi_0^4}{120} \cdot \{S(\pm 3, \pm 3, \pm 3, \pm 2, \pm 1) + \right.\right.$$

$$+ S(\pm 3, \pm 3, \pm 2, \pm 2, \pm 1) + S(\pm 3, \pm 3, \pm 2, \pm 1, \pm 1) +$$

$$+ S(\pm 3, \pm 3, \pm 2, \pm 1, 0) + S(\pm 3, \pm 2, \pm 2, \pm 2, \pm 1) + \tag{A.2.25}$$

$$\pm S(\pm 3, \pm 2, \pm 2, \pm 1, \pm 1) + S(\pm 3, \pm 2, \pm 2, \pm 1, 0) +$$

$$\left.\left.+ \frac{m_0^2}{4} \cdot [S(\pm 3, \pm 4, \pm 3, \pm 2, \pm 1) + S(\pm 3, \pm 2, \pm 3, \pm 2, \pm 1)]\} \right]\! \right]$$

$$\tilde{S}_{\pm 5} = \frac{m_0^5}{32} \frac{\chi_0^4}{120} \cdot S(\pm 5, \pm 4, \pm 3, \pm 2, \pm 1). \tag{A.2.26}$$

Design of the copper vapour laser (laser generator) developed for the display of information of a large screen

Output radiation wavelength λ_1 = 510.6 nm, λ_2 = 578.2 nm

Output average optical power at λ_1	controlled from 0 to 12 W
Output average optical power at λ_2	controlled from 0 to 8 W
Total optical power of radiation *	not less than 20 W
Pulse repetition rate pumping	15.625 kHz
Frequency of repetition of output radiation pulses (frequency of TV lines)	up to 15.625 kHz
Duration of output radiation pulses	15 ns
Divergence of laser radiation	$1.6 \cdot 10^{-4}$ rad
Diameter of the output beam radiation	2 cm
Power supply	~ 220...230 V, 50...60 Hz, 20 A or ~3×190 V, 50...60 Hz, 16 A
Consumed power, no more than	4 kW
Cooling system	industrial water, 1 ... 2.5 1/min

The minimum operating time is determined by the life time of the thyratron (1000 hours) and the active element (2000 hours).

Notes:

* in the beam there can be the wavelengths of light waves λ_1 and λ_2 in any combination.

Control-pumping rack
Dimensions $1130 \times 645 \times 555$ mm^3
Weight   122 kg

Laser emitter unit
Dimensions $1800 \times 380 \times 310$ mm^3
The weight   115 kg

The device and the principle of operation
A simplified block diagram of the laser is shown in Fig. A.3.1.

1. The 'I/O-converter' block is a software-controlled power supply with an output voltage of up to 7 kV and a power of 3.5 kV.

2. In the 'Modulator¦ block there is a laser pump source made on a thyratron commutator.

3. The active laser element can be located in the 'Laser Emitter' block, the 'Optical System' is a control system for output radiation based on acousto-optical modulators and mirrors or acousto-optical deflectors, as well as a non-linear system for converting the radiation of a copper vapour laser into a second harmonic. The 'Control Unit' is used to monitor the operation and maintenance of the laser radiation control system.

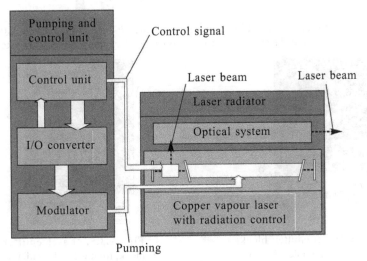

Fig. A.3.1. Block diagram of the laser.

Laser radiator block

The appearance of this block is shown in Fig. A.3.2-A.3.5.

FigureA.3.2 shows the location of the internal units of the 'Laser Emitter' unit.

Figure A.3.6 shows the appearance of the rack 'Pumping and control unit'. Inside the rack there are blocks in accordance with Fig. A.3.7.

The appearance of the blocks with a brief description of the controls and displays is shown in Fig. A.3.8-A.3.13.

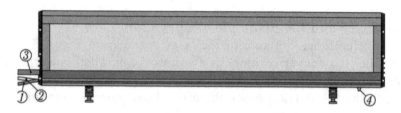

Fig. A.3.2. 'Laser emitter' block, side view: *1* – water inlet; *2* – water outlet; *3* – laser pumping cable; *4* – input of the laser radiation control system

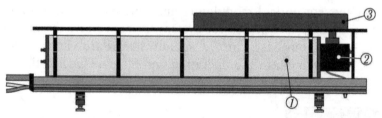

Fig. A.3.3. The 'Laser Emitter' block without a casing: *1* – a cooler with an active element of a copper vapor laser; *2* – laser radiation control system, *3* – optical system.

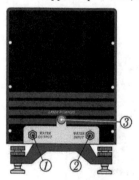

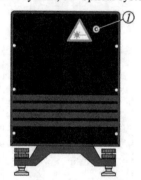

Fig. A.3.4. 'Laser emitter' block, rear view: *1* – water inlet; *2* – water outlet; *3* – laser pumping cable.

Fig. A.3.5. 'Laser emitter¹ block, front view: *1* – radiation output.

Figure A.3.14 shows photographs of the construction of cooled chokes, which are described in Sec. 3.1, and in Fig. A.3.15 and A.3.16 (see color insertion On 11) shows photos of the modulator unit and the high-voltage converter.

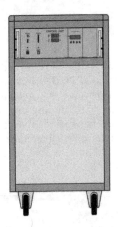

Fig. A.3.6. Pump and control unit stand, front view.

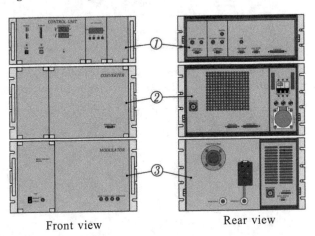

Front view   Rear view

Fig. A.3.7. Arrangement of units in the control-pumping rack: *1* – control unit; *2* – I/O converter block; *3* – modulator unit.

Work with the laser is as follows.

1. The coolant flow is turned on.

2. After pressing the ON button on the control unit, the ON indicator and the A, B, C and W indicators light up. After 10 minutes, the I/O-converter block turns on. During the next 30 minutes, the voltage applied to the modulator increases smoothly. Thus, the active element is heated gradually. This process is displayed using

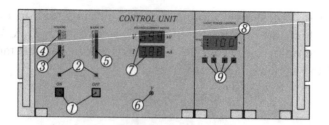

Fig. A.3.8. Front panel of the control unit: 1 – (ON) and (OFF) buttons; 2 – indicators of the mode 'work' (ON) and 'off' (OFF); 3 – indicators of the presence of supply voltages (A, B, C) and coolant flow (W); 4 – indicators of tripping of thyratron disruption; 5 – indicators of feeds and pumping on the active element of a copper vapor laser; 6 – modulator supply voltage regulator; 7 – voltage (V) and current (I) voltage indicators of the modulator. 8 – indicator of the operating mode of the laser radiation control system; 9 – buttons for switching the modes of the laser radiation control system.

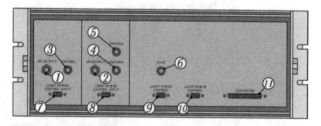

Fig. A.3.9. Rear panel of the control unit: *1, 2* – radio frequency outputs; *3, 4, 5, 6* – control connectors; *7, 8, 9, 10, 11* – connectors for interblock connections

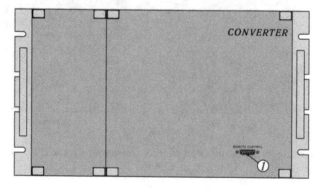

Fig. A.3.10. Front panel of the I/O-converter block: *1* – connector for remote control

the WARM UP indicator bar. Voltage and current at the input of the modulator is displayed with the help of the digital indicators V and I. After reaching the maximum (thus lighting up the F indicator of the WARM UP scale) the voltage value is ~6 kV. The current can change as the active element warms up within 450...550 mA. The

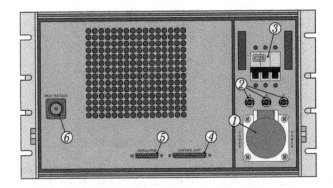

Fig. A.3.11. Back panel of the '???/???-converter". *1* – power connector; *2* – fuses; *3* – batch switch; *4, 5* – connectors for interblock connections; *6* – output high-voltage power connector modulator.

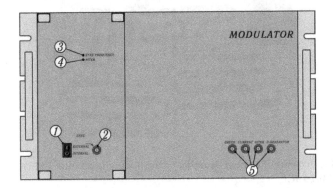

Fig. A.3.12. Front panel of the 'Modulator' block: *1* – external/internal synchronization 1switch; *2* – external synchronization input; *3* – adjustment of the frequency of internal synchronization of a copper vapor laser; *4* – regulation of the thyratron heating voltage; *5* – control connectors: voltage on the thyratron grid **(GREED)**, cathode current of the thyratron **(CURRENT)**, thyratron heating voltage **(HITER)**, voltage on the heater of the hydrogen generator **(H-GENERATOR)**.

total time of the laser output to the operating mode with a nominal output power is 60 minutes from the start-up.

Using the 'Y', 'G', '<", ">' buttons on the front panel of the "Control Unit", the power and the spectral composition of the copper vapor laser radiation are controlled. The '<", ">' buttons set the output power as a percentage of the maximum (indicated on the digital display). The 'Y' (yellow) button turns on/off the yellow component of the radiation, and the 'G' button (green) respectively green. Thus, the spectral composition and the average power of the output radiation are controlled.

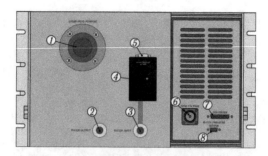

Fig. A.3.13. Rear panel of the 'Modulator' unit: *1* – socket for connecting the pumping cable of the active element of the copper vapour laser; *2* – coolant outlet; *3* – coolant inlet; *4* – the gauge of pressure of a cooling liquid; *6* – input high-voltage power connector modulator; *5, 7, 8* – connectors for interconnects.

Fig. A.3.14. Pictures of the construction of the chokes L_2 (*1*) and L_3 (*2*). *3* – thyratron TGI-2-1000/25 K.

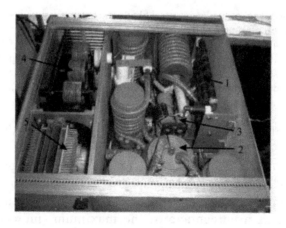

Fig. A.3.15. Photo of a modulator of a copper vapour laser. (1) – choke, (2) - thyratron TGI-2-1000/25 K, (3) – capacitance block C_1, C_2, (4) – charge choke L_0, (5) – voltage converter for heating and the driver of triggering pulses of the thyratron.

Fig. A.3.16. Photograph of the high-voltage converter of a copper vapour laser.

Fig. A.3.17. The appearance of a copper vapour laser.

Fig. A.3.18. Operation of a copper vapour laser.

3. Turn off the laser by pressing the 'OFF' button on the front panel of the control unit. The OFF indicator lights up. After one minute, the laser will automatically turn off.

4. After 30 minutes, turn off the water. Figure A.3.17 presents a photograph of the appearance of the developed laser, and in Fig. A.3.18 a photo of the laser emitter on the side of the output device.

An oscillator–amplifier copper vapour laser system for illumination of architectural structures and formation of vector–graphic images

An original laser system with an output power of up to 50 W at the amplifier output is designed for artistic illumination of historical sights of St. Petersburg in the evening and at night: crosses, spires and domes, as well as for creating vector–graphic images on the walls of buildings. The system is a master oscillator–amplifier system on copper vapour and consists of two identical copper vapour lasers based on active laser elements LT-30 CU assembled in one rack. Below are the technical characteristics of the laser installation.

Specifications

The laser system provides the following technical characteristics:

Wavelength of laser radiation $\lambda_1 = 0.511$ μm, $\lambda_2 = 0.578$ μm.

Average power of laser radiation
at the generator output — 15–18 W,
at the output of the amplifier — 45–50 W.
The duration of laser pulses — 20–30 ns.
Frequency of repetition of laser pulses — 16 kHz.
Divergence of laser radiation — $2\text{–}3 \cdot 10^{-4}$ rad.

Dimensions and weight:

laser emitter	$1800 \times 750 \times 630$ mm³, 200 kg – 1 piece;
power supply and control units	$520 \times 490 \times 350$ mm³, 30 kg – 2 pieces;
optical system	$2600 \times 550 \times 100$ mm3.30 kg.
Total weight of power units and management	290 kg.
Power consumption from 3 phase network	not more than 8 kW. 50 Hz 380 V
Cooling	water, 4–8 l/min.

The time of continuous operation is limited only by the resource of active elements and thyratrons,

Resource of active element and thyratron	> 1000 h.
System availability time	60 min.

The block diagram of the laser installation is shown in Fig. A.4.1 and consists of blocks *1–26* and a system for outputting laser radiation to illuminated objects. Blocks *6* and *16* (high-voltage voltage converter) form a software-controlled power supply with an output voltage of up to 7 kV and a power of up to 4.5 kV (operating voltage 5.2 kV). The converter is made on IGBT transistors according to the scheme of a full bridge with phase control operating at a frequency of 50 kHz.

Pumping of the active elements of the laser LT-30CU is carried out by the modulator blocks *9* and *19*, in which thyratrons of the type TGI-2-1000/25K are used. The control unit for the laser unit *22* is made on the basis of the Microchip microcontroller. With its help, control of the switching on and off of lasers is controlled, emergency modes are monitored, various parameters of the operation of lasers are monitored and set: the active element heating modes (stepwise pump power increase), pumping power, cooling time, thyratron operation modes, the operating time of active elements and thyratrons. It should be noted that the possibility of stabilizing the electric power of the active element supplied to the heating allows to warm up it more efficiently, remove the loads from the excitation system during the transient process of establishing the conductivity of the active element, and also to stabilize the thermal mode of the active element when it reaches a given output power level.

Figure A.4.2 shows the design and location of the units in the rack. The power converter units *6* and *16* are located in separate housings

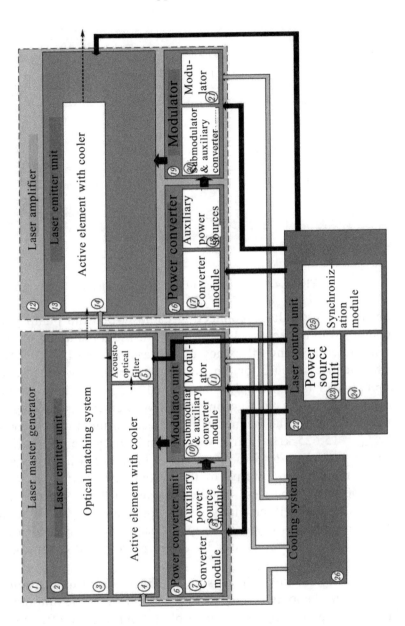

Fig. A.4.1. Block diagram of laser installation.

and connected to the main rack using high-voltage (= 10 kV) and signal cables. As can be seen from Fig. A.4.2, the symmetrical arrangement of active elements and their excitation system in the rack allows one to swap the oscillator and the amplifier. The arrangement of active elements and modulator blocks in one rack made it possible

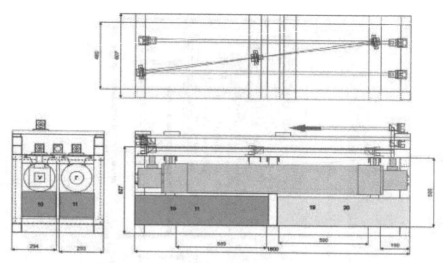

Fig. A.4.2. The design of the laser installation.

to abandon high-voltage connecting cables when the pump pulses were applied to the active element, which led to a reduction in losses and an increase in the efficiency of the laser system by 10–15%.

The electric scheme for the formation of pump pulses of active elements of a laser complex performed according to the known scheme of doubling voltage on the active element using magnetic compression cells. The frequency of the laser installation was chosen to be $f_{rep} = 15.625$ kHz, which was determined by the possibility of its use in the future to form a television image on a large screen.

To increase the reliability of the system, two thyratrons of the TGI-2-1000/25K type were installed in each of the modulator units, which could work either singly with a frequency f_{rep}, or alternately with the frequencies $f_{rep}/2$. The gaps of each of the thyratrons were fed from their glow converter by direct current and controlled by separate submodulators.

Figure A.4.3 shows the design of the modulator unit and its connection to the active element in the rack. Due to the optimization of the discharge circuit, it was possible to increase the efficiency of the AE pump system.

A simplified electrical diagram of the laser power circuits is shown in Fig. A.4.4.

The power of the lasers that make up the laser complex is realized from a three-phase network with a voltage of 3 × 380 V and a frequency of 50 Hz, the phase load is uniformly applied and does not exceed 1.5 kW per phase for one laser. The main sources of

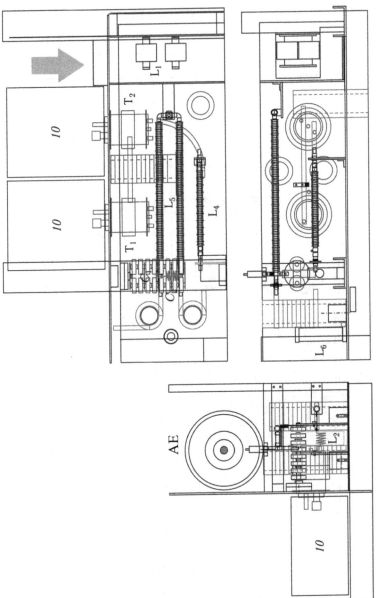

Fig. A.4.3. The modulator unit.

potential interference from the laser are interference from a high-voltage voltage converter made on IGBT transistors in a full-bridge phase-controlled scheme operating at 50 kHz (the main frequencies of possible interference are 100 kHz and 50 kHz), as well as a high-voltage thyratron commutator, impulse pumping on an active laser element (50 ns, 200 A) operating at a frequency of f_{rep}.

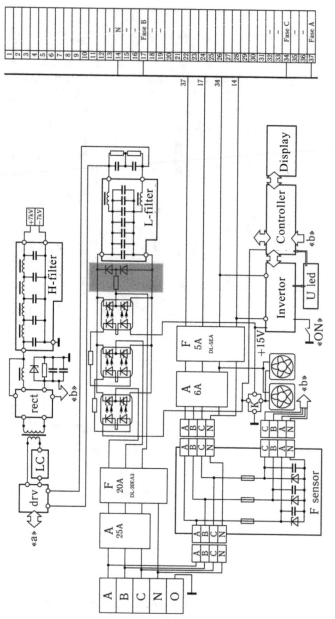

Fig. A.4.4. Simplified electrical circuit of laser power circuits.

The following measures are taken to eliminate the ways of penetration of these interferences into the external environment, as well as into the input electrical circuit.

1. At the input of the power converter units, there is a three-phase noise suppression filter of the type DL-20 EA3.

2. An antisymmetric filter (LV filter) is located at the output of the three-phase rectifier to the power converter.

3. All the auxiliary circuits of the laser are powered by high-frequency converters operating at frequencies of 100 kHz and equipped with input and output filters.

4. High-voltage high-frequency filters are located at the output of the power converter unit and at the input of the modulator unit, which do not allow the impulses of interference from the thyratron commutators to the converter.

5. The heating power and the formation of other auxiliary voltages on the thyratrons are realized from separate high-frequency converters equipped with input and output filters that prevent the penetration of short pump pulses through the thyratron filament into the input electrical network.

6. The active element of the laser has a grounded shielding casing, which prevents air interferences.

7. Air electromagnetic interference from the operating laser is eliminated by optimizing the discharge circuits in the power section of the modulator unit, as well as by shielding possible sources of electromagnetic radiation.

8. All laser units have additional grounded metal housings.

The measures taken have made it possible to reduce the electrical pickups created by the laser complex to the required minimum, which is confirmed by the following factors:

1) control of the laser complex, operating modes and diagnostics of output parameters of lasers is carried out by controllers built on the power supply units of the laser based on the Microchip microcontroller;

2) directly with the lasers is a television receiver with an external antenna that allows you to receive television programs without interference;

3) control and monitoring of individual units of the laser installation (laser radiation control system) is carried out from the computer next to the installation.

Figure A.4.5 shows the optical scheme of a laser installation. The active elements of the generator and amplifier of the laser installation were sealed gas discharge tubes LT-30 Cu were used, the output optical windows of which were made of plane-parallel antireflecting glass plates adhered to the ends of the tube. This made it possible to minimize optical aberrations in the output radiation. The resonator of the master oscillator is made of a three-mirror scheme of a telescopic

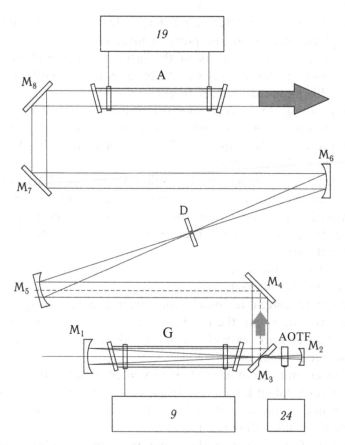

Fig. A.4.5. Optical scheme of laser installation (A – amplifier; G – generator).

unstable resonator with a magnification factor of 30, consisting of spherical concave mirrors M_1, M_2 with radii of curvature R_1 = 3 m and R_2 = 10 cm and a flat mirror M_3 with a communication hole of 1 mm diameter for emission. To control the laser radiation between mirrors M_2 and M_3, an acousto-optically tunable filter (AOTF) made of a paratellurite crystal (TeO_2) is installed inside the resonator.

   AOTF allows to control the amplitude, repetition frequency and wavelength (λ_1 = 0.5106 μm or λ_2 = 0.5782 μm) of laser pulses with high accuracy without changing the mode of heating of the active element and the direction of propagation of laser radiation. To suppress superluminescence in the absence of control pulses, as well as for angular selection, the output radiation from the generator is passed through a spatial filter in the form of diaphragm D, installed in the focus of the mirror telescopic system M_5, M_6. In addition, AOTF from TeO_2 serves as an effective polarizer of laser radiation and replaces the Glan prism. The maximum average power at the output of the generator was 20 W, with a divergence of $2 \div 3 \cdot 10^{-4}$ rad. At the output of the amplifier, the output power was 50 W. To reduce the influence of the amplifier on the output radiation, in the absence of the AOF control signal, as well as to increase the color contrast when switching the wavelength, an additional delay between the pump and generator pumping pulses was introduced.

The control unit *24* made it possible to control the appearance time of each light pulse, and the amplitude of the light pulses could change by changing the amplitude of the radio pulses at the input of the AOF. Thus, it was possible to safely aim at the objects of illumination and to adjust the optical guidance system at a low level of light power.

The output radiation from the output of the amplifier sequentially passed through the beam splitters and was divided into 15 beams, each of which was directed to its illuminated object. The power of each ray was selected proportionally to the distance to the object. The size of the light spot on the objects was adjusted using lens telescopes installed in each of the 15 light channels.

The created laser complex is installed on the roof of the Grand Palace in the center of St. Petersburg and is used to illuminate at night crosses of Orthodox churches and the main symbols of the city – a boat at the Admiralty and an angel on the spire of the Peter and Paul Fortress. The farthest object – Smolny Cathedral – was at a distance of 3.5 km.

Figures A.4.6 and A.4.7 show photographs of the operating installation and the units (*6*) and (*16*) (see Fig. A.4.1) of high-voltage converters, and Figs. A.4.8 and A.4.9 are photographs of the objects to be illuminated.

Fig. A.4.6. Working installation.

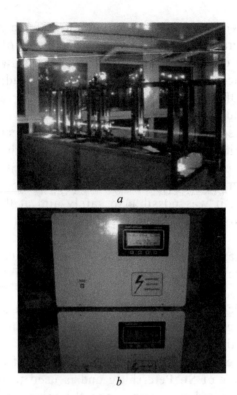

a

b

Fig. A.4.7. Photograph of the working installation (*a*) and high-voltage converter units (*b*).

Fig. A.4.8. Illuminated objects.

References

1.   Zvorykin V.K., Morton D.A., Television. – Moscow: IL, 1956. – 780 p.
2.   Vasiliev A.A., et al., . Spatial light modulators. – M.: Radio i Svyaz', 1987. – 320 p.
3.   Mukhin I.A., Development of liquid crystal monitors. Part 1. BROADCASTING Television and radio broadcasting. – 2005. – V. 46, No. 2. – P. 55–56.
4.   Mukhin I.A., Development of liquid crystal monitors. Part 2. BROADCASTING Television and radio broadcasting. – 2005. – V. 48, No. 4. – P. 71–73.
5.   Mukhin I.A., Principles of image sweep and modulation of the luminescence brightness of a cell of a plasma panel. Trudy uchebnykh zavedenii svyazi. SPbGUT. – 2002. – No. 168. – P. 134–140.
6.   Friend R.H., et al., Nature. – 1999. – Vol. 397. P. 121–128.
7.   Mayskaya V., Organic light-emitting diodes. Electronika: NTB. – 2007. – No. 5. – P. 39–46.
8.   Sinitsyn N.I., et al., Radiotekhnika. 2005. – No. 4. – P. 35–40.
9.   Canon, Toshiba Bring SED Panels to Reality. Display Devices Fall. – 2004. – P. 35.
10.   Ulasyuk V.N., Quantum scopes. – Moscow: Radio i Svyaz', 1988. – 256 p.
11.   Mokienko O.M., Electronika: NTB. – 2000. – No. 6. – P. 54–56.
12.   Robinson D.M., The supersonic light controls and its application to television with special reference to the Scophony television receiver. Proceedings of the IRE. – 1939. – Vol. 27, No. 8. – P. 483–487.
13.   Wikkenhauser G., Synchronization of Scophony television receiver. Proceedings of the IRE. – 1939. – Vol. 27, No. 8. – P. 492–496.
14.   Korpel A., et al., A television display using acoustic deflection and modulation of coherent light. Proceedings of the IEEE. – 1966. – Vol. 54, No. 10. P. 1429–1437.
15.   Yamamoto M., Taneda T., Laser imaging devices. Achievements in the technique of image transmission and reproduction, vol. 2: Trans. from English. – Moscow: Mir, 1979. – 286 p.
16.   Gordon E.I., A review of acousto-optical deflection and modulation devices. Proceedings of the IEEE. – 1966. – Vol. 54, No. 10. P. 1391–1401.
17.   Aksenov E.T., et al. Kvantovaya elektronika. Trudy LPI – 1974. – No. 366. – P. 69–74.
18.   Taneda T., et al. High quality laser television display. Journal of the SMPTE. – 1973. – No. 6.
19.   Gorod J., Knox J. D., Goedertier P. V. A television-rate laser scanner. RCA Review.

– 1972. – V. 33, No. 12. – P. 623–674.

20. Geoffrey G. F. An experimental laser-photo chromic display system. The Radio and Electronic Engineer. – 1970. – V. 39, No. 3. – P. 123–129.

21. Benedichuk I.V.,et al., Tekhnika kino i televideniya. – 1978. – No. 6. – P. 3–10.

22. Watson W.N., Korpel A., Appl. Opt – 1970. – Vol. 9, No. 5. – P. 1176–1179.

23. Klima M., Slaboproudy obzor. – 1979. – V. 40, No. 9. – P. 415–421.

24. Yamada Y., Yamamoto M., Nomura S., Large screen laser color TV projector. Proc. Int. Quantum Electron., 6th, Kyoto. – 1970 – P. 242–243.

25. Yamamoto M., A 1125-scanning-line laser color TV display. Hitachi Rev. – 1975. – No. 24. P. 89–94.

26. Nowicki T., Electro-Opt. System Design – 1974. – Vol. 6, No. 2. – P. 23–28.

27. Okolicsanyi F.,Wireless Eng. – 1937 – Vol. 14. – P. 527–536.

28. Bergmann L., Ultrasonics. – New York: J. Wiley. – 1938. – P. 58–63.

29. U.S. Patent No. 3818129. Laser imaging device / M. Ymamoto. – June 18, 1974.

30. Damon R., et al., Interaction of light with ultrasound: The phenomenon and Its Application. Physical Acoustics, V. 7: translated from English / Ed. W. Mason and R. Thurston. – M.: Mir. – 1974. – P. 311–426.

31. Lowry J.B., et al., Optics and Laser Technology. – 1988 – Vol. 20, No. 5. – P. 255–258.

32. Mokrushin Yu. M. The acousto-optical system for mapping information with a pulsed laser on copper vapors: Author's abstract of Ph.D thesis – Leningrad.: LPI, 1987. – 16 p.

33. Mokrushin Yu.M., Shakin O.V., Journal of Russian Laser Research. – New York. – 1996. – Vol. 17, No. 4. – P. 381–393.

34. Martinsen R.J., Aylward R.P., Photonics Spectra. – 1996. – No. 11. – P. 109–114.

35. Martinsen R.J., et al., Pulsed RGB laser for large screen video displays. SPIE Proc. 3000–30. – P. 150–160.

36. Laser Focus World, May 1999, p. 13.

37. Armstrong J.A., et al., Phys. Rev. 127, 1918 (1962).

38. Franken P.A. and Ward H.F., Rev. Mod, Phys. 35, P. 23–39 (1963).

39. Lim E.J., et al., *Electronics Letters* 25 (3), 174–5 (1989).

40. Dmitriev V.G., Tarasov L.V. Applied nonlinear optics. – M.: Fizmatlit, 2004 – 512 pp.

41. Hu X.P., Optics Letters. – 2008. – Vol. 33, No. 4. – P. 408–410.

42. U.S. Patent, No 5,828,424. Process and apparatus for generation at least three laser beams of different wavelength for the display of color video pictures / R. Wallenstein. – 1998.

43. Nebel A., Ruffing B., Wallenstein R., A 19 W RGB solid-state laser source for large frame laserprojection displays. Laser and Electro-Optics Society Annual Meeting. LEOS IEEE. – 1998. – P. 395–396.

44. Nebel A., et al., Laser Focus World. – 1999. – P. 263–266.

45. Brunner F., et al., Optics Letters. – 2004. – Vol. 29, No. 16. – P. 1921–1923.

46. Innerhofer E., et al., J. Opt. Soc. Am. B. – Vol. 23, No. 2. – P. 265–274.

47. Watson J.P. et al., Proc. SPIE. – 2004. – Vol. 5364. – P. 116.

48. Shchegrov A.V.. Proc. SPIE. – 2004. – Vol. 5332. – P. 151.

49. Solgaard O., et al., Optics Letters. – 1992. – Vol. 17, No. 9. – P. 688–690.

50. Bloom D.M., Proc. SPIE, Projection Displays III. – 1997. – Vol. 3013. – P. 165–171.

51. Trisnadi J.I., et al., Overviewand Applications of Grating Light Valve Based Optical WriteEngines for High-Speed Digital Imaging. Proc. Micromachining and Microfabrication Symp., Photonics West. – San Jose, CA, Jan. 26, 2004.

52. Petrash G.G., Usp. Fiz. Nauk. – 1971. – Vol. 105, No. 4. – P. 645–676.

53. Little C.E., Metal Vapour Laser: Physics, Engineering and Applications. – Chichester (UK): J.Wiley and Sons, 1999. – 620 p.

54. Lyabin N.A., et al., Kvant. elektronika. – 2001. – V 31, No. 3. – P. 192–202.

55. Grigor'yants A.G., et al., Copper vapor lasers: design, characteristics and applications. – Moscow: Fizmatlit 2005. – 312 p.

56. Bokhan P.A., et al., Kvant. elektronika – 1980. – V. 7, No. 6. – P.1264–1269.

57. Isaev A.A., et al., Kvant. elektronika. – 1983. – V. 10, No. 6. – P. 1183–1189.

58. Isaev A.A., et al., Kvant. elektronika. – 1977. – V. 4, No. 7. – P. 1413–1417.

59. Markova S.V., et al., Kvant. elektronika. – 1978. – V. 5, No. 7. – P. 1585–1587.

60. Markova S.V., Cherezov V.M., Kvant. elektronika. – 1977. – V. 4, No. 3. – P. 614–619.

61. Divin V.D., Isakov V.K., Sov. J. Quantum Electron. – 1986. – V. 16, No. 8. – P. 1081–1085.

62. Markova S.V., et al., – 1977. – V. 4, No. 5. – P. 1154–1155.

63. Linevsky M.J., Karrus T.W., Appl. Phys. Lett. – 1978. – V. 33, No. E8. – P. 720–721.

64. Isaev A.A., et al., Trudy FIAN – 1987. – V. 181. – P. 3–17.

65. Isaev A.A., et al., Kvant. elektronika. – 1976. – V. 3, No. 8. – P. 1802–1805.

66. Bokhan P.A., Gerasimov V.A., Kvant. elektronika. – 1979. – V. 6, No. 3. – P. 451–455.

67. Lewis R.R., et al., Laser focus. – 1988. – V. 24, No. 4. – P. 92, 94–96.

68. Soldatov A.N., Fedorov V.F., Izv. VUZ. Fizika – 1983. – No. 9. – P. 80.

69. Kalugin M.M., et al., Pis'ma Zh. Teor. Fiz. – 1980. – V. 6, No. 5. – P. 280–283.

70. Isaev A.A., Kazaryan M.A., Kvant. elektronika..– 1977. – V. 4, No. 2. – P. 451–453.

71. Zubov V.V., et al., Kvant. elektronika.– 1983. – V. 10, No. 9. – P. 1908–1910.

72. Gradoboev Yu.G., et al., Pribory i tekhnika eksperimenta. – 1990. – No. 6. – P. 118–120.

73. Grekhov I.V., et al., Journal of Russian Laser Research. New York. – 1996. – V. 17, No. 4. – P. 362–364.

74. Zemskov K.I., et al. Kvant. elektronika. – 1976. – V. 3, No. 1. – P. 35–43.

75. Zemskov K.I., et al., Kvant. elektronika. – 1974. – V. 1, No. 4. – P. 863–869.

76. Belyaev V.P., et al., Elektronnaya promyshlennost'. – 1984. – V. 10, No. 138. – P. 28–30.

77. Du C., et al., Optics Express. – 2005. – V. 13, No. 21. – P. 8591–8595.

78. Du C., et al., Optics Express. – 2005. – V. 13, No. 6. – P. 2013–2018.

79. Haiyong Z., et al., Optics Express. – 2008. – V. 16, No. 5. – P. 2989–2994.

80. Konno S., et al., Optics Letters. – 2000. – V. 25, No. 2. – P. 105–107.

81. Bo Y., et al., Applied Optics. – 2006. – V. 45, No. 11. – P. 2499–2503.

82. Tang H., et al., Chinese Optics Letters. – 2009. – Vol. 7, No. 9. – P. 812–814.

83. Lee D., Moulton P.F., High-efficiency, high-power, OPO-based RGB source. Conference on Lasers and Electro-Optics (CLEO) 2001. – Vol. 56 of OSA Trends in

Optics and Photonics Series (Optical Society of America, 2001), paper CThJ2, p. 424.

84. Moulton P.F., et al., High-power RGB Laser Source for displays. Presented at the IMAGE 2002 Conference Scottsdale. Arizona 8–12 July. – 2002.

85. Dixon R.W., IEEE J. Quantum. Electron. – 1967. – V. QE-3, No. 2. – P. 85–93.

86. Korpel A., Acousto-optics.Applied Solid State Science, Advances in Materials and Device Research. – 1972. – V. 3, No. 2. – P. 71–80.

87. Jieping X., Stroud R., Acousto-Optic Devices: Principles, Design and Applications. – John Wiley & Sons, Inc., 1992. – 652 p.

88. Gordon E.I., Proc. IEEE. – 1966. – V. 54, No. 10. – P. 1391–1401.

89. Chang I.C., IEEE Trans. Son. Ultrason. – 1976. – V. SU-23, No. 1. – P. 2–22.

90. Goutzoulis A., et al., Design and fabrication of acoustooptic devices. – Marcel Dekker Inc.: N. York, 1994. – 497 p.

91. Mustell R., Parygin V.N. Methods of modulation and scanning of light. – Moscow: Nauka, 1970. – 295 p.

92. Rebrin Yu.K., Control of the optical beam in space. – Moscow: Soviet radio, 1977. – 336 p.

93. Magdich L.N., Molchanov V.Ya., Acoustooptical devices and their application. – Moscow: Soviet radio, 1978. – 111 p.

94. Balakshii V.I., et al., Physical principles of acoustooptics. – M.: Radio i svyaz', 1985. – 280 p.

95. Kulakov S.V., Acoustooptical devices for spectral and correlation analysis of signals. – Leningrad: Nauka, 1978. – 144 p.

96. Rodes W.T., Acousto-optical signal processing. Convolution and correlations. TIIER. – 1981. – V. 69, No. 1. – P. 74–91.

97. Rodes W.T., Architecture of acousto-optic algebraic processors, TIIER. – 1984. – V. 72, No. 7. – P. 80–91.

98. Psaltis, D., Two-dimensional optical signal processing using one-dimensional input devices, TIIER. – 1984. – V. 72, No. 7. – P. 240–255.

99. Harris S.E., Wallace R.W., J. Opt. Soc. Amer. – 1969. – Vol. 59, No. 6. – P. 744–747.

100. Chang I.C., Proc. SPIE. – 1976. – V. 90. – P. 12–22.

101. Yano T., Watanabe A., J. Appl. Optics. – 1976. – Vol. 15, No. 9. – P. 2250–2258.

102. Voloshinov V.V., et al., Proc. SPIE. – 2001. – Vol. 4353. – P. 17–22.

103. Balakshii V.I., et al., Kvant. elektronika. – 1979. – V. 6, No. 5. – P.965-971.

104. Warner A.W., et al., J. Appl. Phys. Letts. – 1972. – Vol. 43, No. 11. – P. 4489–449.

105. Yano T., et al., J. Appl. Phys. Letts. – 1975. – Vol. 26, No. 12. – P. 689–691.

106. Kludzin V.V., Presleyev L.N., Coherent storage of radio impulses in acousto-optic delay lines. Acoustooptical methods and information processing technique. Intercollegiate collection. – No. 142. – Leningrad: LETI, 1980. – P. 75–81.

107. Bakinovskii K.I., et al., Priboty i tekhnika eksperimenta. – 1986. – No. 3. – P. 247.

108. Nelson D.F., Lax M., Phys. Rev. – 1971. – V. B3. – No. 8. – P. 2778–2794.

109. Nai J. Physical Properties of Crystals. – Moscow: Mir, 1967.

110. Lemanov V.V., Shakin O.V., Fiz. Tverdogo Tela – 1972. – Vol. 14, No. 1. – P. 229–236.

111. IRE Standards on Piezoelectric Crystals. Proc. IRE. – 1949. – V. 37. – P. 1378; 1957. – V. 45. – P. 354; 1958. – V. 46. – P. 765; 1961. – V. 49. – P. 1162.

112. Berlinkur D., et al., in: Physical acoustics / Ed. W. Mason. – M.: 1966. – V. 1A. – P. 204.

113. Ivanovskii G.F., Petrov V.I., Ion-plasma treatment of materials. – Moscow: Radio i svyaz', 1986. – 232 p.

114. Author Cert. 1529785 (USSR). Method for the preparation of paratellurite single crystals / O.V., et al.. – 1989.

115. Bond V.L ,Technology of Crystals. – Moscow: Nedra, 1980. – 304 p.

116. Author Cert. 145670 (USSR). Precision orientation of crystals. / O.V. Shakin, et al., – 1980.

117. Author Cert. 1256467 (USSR). Method for the production of an acousto-optical monocrystal NaBi $(MoO_4)_2$ / O.V. Shakin, et al. – 1986.

118. Author Cert. 1354789 (USSR). The method of obtaining acousto-optical mono-crystals of NaBi $(MoO_4)_2$ / O.V. Shakin, S. V. Akimov, S.Yu. Yermakov, TM Stolpa-kova, A. N. Grishmanovsky. – 1987.

119. Shutilov V.A., Fundamentals of Ultrasound Physics. Tutorial. – Leningrad: Lenin-grad University Press, 1980. – 280 p.

120. Physical Values. Reference book / Babichev A.P., et al., – Moscow: Energoatomiz-dat, 1991. – 1232 p.

121. Pape D.R., Opt. Eng. – 1992. – V. 31, No. 10. – P. 2148–2158.

122. Acoustic crystals. Handbook. Ed. M.P. Shaskolskaya. – Moscow: Nauka, 1982. – 632 p.

123. Uchida N., Ohmachi Y., J. Appl. Phys. – 1969. – Vol. 40, No. 12. P.4692–4695.

124. Kuzin A.G., Investigation of the influence of elastic and optical anisotropy of a me-dium on the parameters of acoustooptical control devices for laser beams: Disserta-tion thesis. – Leningrad: LIAP, 1979. – 264 p.

125. Turok I.I., Golovei M.M., Growing paratellurite single crystals and the temperature dependence of their refractive indices. II All-Union Conference "Actual problems of obtaining and using ferroelectric and piezoelectric materials": – M., 1984. – P. 264.

126. Proklov V.V., Radiotekhnika i elektronika. – 1980. – V. 25, No. 7. – P. 1543–1545.

127. Molotok V.V., Razzhivin B.P., Influence of attenuation of acoustic waves on the characteristics of acoustooptical spectrum analyzers. Acoustooptical methods and information processing technique. Intercollegiate collection, vol. 142. – Leningrad: LETI, 1980. – P. 10–15.

128. Pisarevskii Yu.V., Silvestrova I.M., Kristallografiya. – 1973. – V. 19, No. 5. – P. 1003–1013.

129. Uchida N., Ohmachi P., Japan. J. Appl. Phys. – 1970. – V. 9, No. 1. – P. 155–156.

130. Bogdanov S.V., Bolshevova T.A., Avtometriya. – 1985. – No. 5. – P. 34–41.

131. Warner A.W., et al., J. Appl. Phys. Lett. – 1972. – V. 43, No. 11. – P. 4489–4495.

132. Yano T., et al., J. Appl. Phys. Lett. – 1975. – V. 26, No. 12. – P. 689–691.

133. Brillouin L., La diffraction de la lumiere par des ultrasons, Act. Sci. Ind. – 1933. – V. 59. – P. 1–31.

134. Raman C.V., Nath N.S.N., The diffraction of light by high frequency sound waves. Part I. Proc. Ind. Acad. Sci. – 1935. – V. 2A. – P. 406–412.

135. Raman C.V., Nath N.S.N., The diffraction of light by high frequency sound waves. Part II. Proc. Ind. Acad. Sci. – 1935. – V. 2A. – P. 413–420.

136. Raman C.V., Nath N.S.N., The diffraction of light by high frequency sound waves. Part III. Proc. Ind. Acad. Sci. – 1936. – V. 3A. P. 75–84.

137. Raman C.V., Nath N.S.N., The diffraction of light by high frequency sound waves. Part IV. Proc. Ind. Acad. Sci. – 1936. – V. 3A. – P. 119–125.

138. Raman C.V., Nath N.S.N., The diffraction of light by high frequency sound waves. Part V. Proc. Ind. Acad. Sci. – 1936. – V. 3A. – P. 459–469.

139. Rytov S.M., Izv. AN SSSR. Ser. fiz. – 1937. – No. 2. – P. 223–259.

140. Quate C.F., et al., Proc. IEEE. – 1965. – Vol. 53, No. 10. – P. 1604–1623.

141. Born M., Wolf E. Fundamentals of Optics. – Moscow: Nauka, 1970. – 855 p.

142. Soroka V. V., Akusticheskii zhurnal. – 1973. – V. 19. – P. 877–884.

143. Markuse D., Optical waveguides. – Moscow: Mir, 1974. – 576 p.

144. Parygin V.N., Chirkov L.E., Radiotekhnika i elektronika. – 1973. – V. 18, No. 4. – P. 703–712.

145. Parygin V.N., Chirkov L.E., Radiotekhnika i elektronika. – 1974. – V. 19, No. 6. – P. 1178– 1186.

146. Parygin V.N., Chirkov L.E., Kvant. elektronika. – 1975. – V. 2, No. 2. – P. 318–326.

147. Martynov A.M., Radiotekhnika i elektronika. – 1977. – V. 22, No. 3. – P. 533–540.

148. Stashkevich A.A., Optika i spektroskopiya. 1978. – V. 45, No. 5. – P. 967–973.

149. Kulak G.V., Zhurnal tekhnicheskoi fiziki. – 1997. – V. 67, No. 9. – P. 80–82.

150. Kulak G.V., Nikolaenko T.V., Zhurnal prikladnoi spektroskopii – 2006. – V. 73, No. 6. – P. 819–823.

151. Mikhailovskaya A.S., Mikhailovskaya L.V., Optika i spektroskopiya. – 2011. – V. 110. – No. 2. – P. 317–323.

152. Petrun'kin V.Yu., Vodovatov I.A., Izv. VUZ. Radiofizika. – 1983. – V. 26. – No. 12. – P. 1570–1578.

153. Petrunkin V.Yu., et al., in: Processing of radio signals by acousto-electronic and acousto-optic devices. Proceedings. – Leningrad: Nauka, 1983. – P. 51–59.

154. Petrukin V.Yu., et al., Izv. VUZ. Radiofizika.. – 1983. – V. 26. – No. 8. – P. 1021– 1029.

155. Petrukin V.Yu., Vodovatov I.A. Izv. VUZ. Radiofizika. – 1984. – V. 27. – P. 332–340.

156. Agranovich V.M., Ginzburg V.L., Crystal optics with allowance for spatial dispersion and the theory of excitons. – Moscow: Nauka, 1979. – 432 p.

157. Sirotin Yu.I., Shaskol'skaya M.P., Fundamentals of Crystallophysics. – Moscow: Nauka, 1975. – 680 p.

158. Vainshtein L.A., Electromagnetic waves. – Moscow: Sov. Radio, 1957. – 581 p.

159. Madelung E., Mathematical apparatus of physics. – M: Fizmatgiz, 1960. – 618 p.

160. Smirnov V.I., Course of Higher Mathematics. V. 4. – Moscow: GIT-TL, 1957. – 812 p.

161. Klein W.R., Cook B.D., Unified approach to ultrasonic light diffraction. IEEE Trans. Son. Ultrason. – V. SU-14. – No. 3. – P. 123–134.

162. Papulis A. The theory of systems and transformations in optics. – Moscow: Nauka, 1971. – 495 p.

163. Vinogradova M.B., et al., Theory of waves. – Moscow: Nauka, 1979. – 383 p.

164. Shmakov P.V., Television. – Moscow: Svyaz', 1979. – 432 p.

165. Randolph J., Morrison J., Appl. Opt. – 1971. – V. 10, No. 6. – P. 1453–1454.

166. Pilipovich V.A., Shcherbak Yu.M., Vestn. AN BSSR, Ser. Fiz. Mat. – 1975. – No. 4. – P. 100–104.

167. Muchnik M.L., et al., Pribory i tekhnika eksperimenta – 1983. – No. 3. – P. 93–94.

168. Belyaev V.P., et al., Kvant. elektronika. – 1985. – V. 12, No. 1. – P. 74–79.

169. Isaev A.A., Trudy FIAN. – 1987. – V. 181. – P. 35–53.

170. Kazarian M.A., et al., Copper vapor laser with intra-cavity acousto-optic output control. XX International Quantum Electronics Conference. Technical Digest. – Sydney, Australia, 14–19 July 1996.

171. Kazarian M.A., et al., Physica Scripta. – 1996. – V. 49. – P.108–110.

172. Kazarian M.A., et al., Copper Vapor Laser with Intra-Cavity Acousto-Optic Output Control. EOS Topical Meeting Digest Series: Vol. 15. Advances in Acousto-Optics. – St. Petersburg, Russia, June 24–25 1997. – P. 108–110.

173. Chang J.J., Opt. Lett. – 1995. – V. 20, No. 6. P. 575–577.

174. Moncorge R., et al., et al. IEEE J. Quantum Electron. – 1988. – V. 24, No. 6. – P. 1049–1051.

175. Bartoshevich S.G., et al., Kvant. elektronika.. – 1989. – V. 16, No. 2. – P. 212–217.

176. Kruzhalov S.V., et al., Pis'ma v Zh. Teor. Fiz. – 1999. – V. 25, no. 18. – P. 12–17.

177. Kazaryan M.A., et al., Kvant. elektronika. – 1998. – V. 25, No. 9. – P. 751–752.

178. Kane D.M., J. Applied Optics. – 1994. – V. 33, No. 18. – P. 3849–3856.

179. Sancher A., et al., IEEE Journ. of QE. – 1988. – V. 24, No. 6. – P.995–1002.

180. Basiev T.T., et al., Kvant. elektronika. – 1997. – V. 24, No. 7. – P. 591–595.

181. Ustida N., Materials and methods of acousto-optic deflection. TIIER. – 1973. – V. 61, No. 8. – P. 21–43.

182. Author Cert. No. 1127440 (USSR). Acoustooptical deflector from paratellurite (its variants) / A.G. Kuzin, Yu.M. Mokrushin. – 1983.

183. Fedorov F.I., Theory of elastic waves in crystals. – Moscow: Nauka. – 1965. – 386 p.

184. Arlt G., et al., Elastic and piezoelastic properties of paratellurite. The 6th International congress on acoustic. – Tokio. – 1968. – P. 89–92.

185. Ohmachi Y., Uchida N., et al., J. Appl. Phys. – 1970. – V. 41. P. 2307–2311.

186. London S.E., Tomashevich S.V., Handbook on high-frequency transformer devices. – M.: Radio i svyaz'. – 1984. – 216 p.

187. Kryzhanovskiy V.D., Kostykov Yu.V., Color and black and white television. – M.: Svyaz'. – 1980. – 336 p.

188. Patent of the Russian Federation No. 2104617. Laser projection system for displaying television information (variants). Yu.M. Mokrushin, O.V. Shakin. – Publ. in BM – 1995. No. 28.

189. Kuzin A.G., et al., Information display device based on a copper vapor laser. Proc. doc. 4th All-Union Conf. :Optics of Lasers. – Leningrad: GOI. – 1984.– P. 352.

190. Kuzin A.G., et al., Optoelectronic method for the formation of TV images. Theses of the report of the All-Union Conf. 'Development and improvement of television technology.' – M.: Radio i svyaz', 1984. – P. 92.

191. Kuzin A.G., et al., Acoustooptical device for displaying and recording information. Abstracts of the II All-Union Conf. 'Formation of the optical image and methods of its processing", vol. 2. – Chisinau. – 1985. – P. 7.

192. Mokrushin Yu.M., et al., Display of television information on a large screen using pulsed copper vapor. Proc. III All-Union Conf. 'The use of lasers in technology and systems of information transmission and processing.' – Tallinn. – 1987. – V. 3. – P.140–142.

193. Petrov M.P., et al., Photosensitive media in holography and optical processing of information. – Leningrad: Nauka. – 1983. – 270 s.

194. Gradoboev Yu.G., et al., The image formation on the PRIZ software with the help of an acousto-optical system, Proc. 6th All-Union School-Seminar on Optical Processing of Information, vol. 2. – Frunze. – 1986. – P. 108–109.

195. Gradoboev Yu.G., et al., Special features of pulse recording information on PRIZE software. Space-Time Modulators of Light for Optical Information Processing: Sat. articles. – Leningrad. – 1987. – P. 64–73.

196. Miroshnikov M.M., Theoretical Foundations of Optoelectronic Devices. – Leningrad: Mashinostroenie. 1983. – 696 s.

197. Silverstova I.M., et al., Crystallography. – 1975. – V. 20. – P. 1062.

198. Vodovatov I.A., in: Acousto=optical Devices of Radioelectronic Systems. Leningrad: Nauka. – 1988. – P. 98–106.

199. Vasil'ev Yu.P., et al., Svetotekhnika. – 1998. – No. 5. – P. 7–10.

200. Kazaryan M.A., et al., Izv. AN. Ser. fiz. – 1999. – V. 63, No. 6. – P. 1190–1191.

201. Gulyaev Yu.V,, et al., Laser Physics. – 2002. – V. 12, No. 8. – P. 6–18.

202. Gradoboev Yu. G., et al., Kvant. elektronika. – 2004. – V. 34, No. 12. – P.1133–1137.

203. Gulyaev Yu. V., et al., Laser in engineering. – 2005. – V. 15, No. 5–6. – P. 293–311.

204. Gulyaev Yu.V., et al., Stroit. Mater. Oborud. Tekhnol. 21 veka. 2007. – No. 1. – P. 51.

205. Mokrushin Yu.M., – No. 2. – P. 93–105.

206. Mokrushin Yu.M., Nauchno-tekhn. vedomosti SPbGPU. Fiz. Matem. Nauki.– 2011. – No. 3. – P. 99–109.

207. Mokrushin Yu.M., Nauchno-tekhn. vedomosti SPbGPU. Fiz. Matem. Nauki.– 2011. – 2011. – No. 4. – P. 118–129

208. Shakin O.V., et al., Kratkie soobshcheniya po fizike. FIAN – 2011. – No. 8. – P. 21–29.

209. Physical Values. Reference book / Babichev A.P., et al., – Moscow: Energoatomizdat, 1991. – 1232 p.

210. Yariv A., Yukh P., Optical waves in crystals. – M.: Mir, 1987. – 616 p.

211. Kulakov S.V., Ultrasonics Symposium, 1994. Proc. IEEE Vol. 2, Issue 1–4, Nov. 1994, P. 851–854.

212. Hanuza J., et al., J. Raman Spectr. – 1999. – V. 28. – P. 953.

213. Kostova E., Kostov M., Appl. Optics. – 1985. – V. 24. – P. 1726

214. Akimov S.V., et al., Solid State Physics (Russ.). – 1977. – V. 19. – P. 1832.

215. Narasimhamurti T.S.. Photoelastic and electro-optic properties of cryostats. – Moscow: Mir, 1984. – 624 p.

216. Uchida N., Phys. Rev B. – 1971. – V. 4. – P. 3736.

217. Korn D.M., e1t al., Phys. Rev. B., – 1973. – V. 8. – P. 768.

218. Gulyaev Yu.V., et al., Kvant. elektronika. – 2015. – V. 45. – No. 4. – P. 283–300.

Index

A

acousto-optical deflectors (AOD) 5, 6, 7, 62, 63, 64, 65, 66
acousto-optical modulators 4, 6, 16, 155, 161, 163, 164, 214, 215, 216
acousto-optical water deflector 4
angle
  Bragg angle 9, 138, 157, 213
  Brewster angle 186
  cut-off angle 43
AOM (acousto-optical modulators) 4, 5, 6, 12, 15, 16, 22, 55, 58, 59,
      60, 65, 66, 67, 86, 87, 88, 89, 90, 93, 94, 95, 97, 98, 99, 106,
      108, 109, 110, 119, 136, 154, 155, 157, 160, 163, 164, 165, 166,
      167, 168, 169, 172, 174, 175, 176, 178, 179, 180, 182, 184, 185,
      188, 189, 190, 191, 193, 194, 195, 196, 197, 200, 203, 207, 208,
      209, 210, 211, 212, 213, 215

B

backscattering 27, 29, 32, 34, 84
Barco Reality 812 2
Bragg diffraction 27, 56, 57

C

cell
  acousto-optical cell 9, 35, 37, 40, 43, 47, 48, 49, 50, 51, 54
  Kerr cell 8
coefficient
  electromechanical coupling coefficient 38, 40, 41, 43
criterion
  Rayleigh criterion 136, 191, 197, 199, 207, 212, 214
crystal
  $KTiOAsO_4$ 11
  LBO crystal 21, 22
  LiB_3O_5 11
  lithium niobate crystals 217, 218
  molybdate crystal 51, 51, 52, 221